BULLETIN OF THE OHIO BIOLOGICAL SURVEY
New Series

Volume 12 Number 3

AF894025

Revision of the Nearctic Species of the Genus *Polypedilum* Kieffer (Diptera: Chironomidae) in the Subgenera *P.* (*Polypedilum*) Kieffer and *P.* (*Uresipedilum*) Oyewo and Sæther

David E. Maschwitz
and
Edwin F. Cook

Published by
Ohio Biological Survey
College of Biological Sciences
The Ohio State University

Columbus, Ohio 43212-1192 2000

Ohio Biological Survey

Bulletin New Series: ISSN: 0078-3994
Volume 12 Number 3 ISBN: 0-86727-130-2
Library of Congress Number: 98-066937

Director
Brian J. Armitage

Editor
Veda M. Cafazzo

Editorial Committee

Barbara K. Andreas, Ph. D., Cuyahoga Community College & Kent State University
Brian J. Armitage, Ph. D., Ohio Biological Survey
Veda M. Cafazzo, M.S., Ohio Biological Survey
Benjamin A. Foote, Ph. D., Kent State University (Emeritus)
Jane L. Forsyth, Ph. D., Bowling Green State University (Emeritus)
Eric H. Metzler, B.S., The Ohio Lepidopterists
Scott M. Moody, Ph. D., Ohio University
David H. Stansbery, Ph. D., The Ohio State University (Emeritus)
Ronald L. Stuckey, Ph. D., The Ohio State University (Emeritus)
Elliot J. Tramer, Ph. D., The University of Toledo

Literature Citation

Maschwitz, David E. and Edwin F. Cook. 2000. Revision of the Nearctic Species of the Genus *Polypedilum* Kieffer (Diptera: Chironomidae) in the Subgenera *P.* (*Polypedilum*) Kieffer and *P.* (*Uresipedilum*) Oyewo and Sæther. Ohio Biological Survey Bulletin New Series Volume 12 Number 3. vii + 135 p.

Cover Photograph
Top - *Polypedilum braseniae* larva on *Nymphaea odorata* leaf; Bottom - *Polypedilum braseniae* pupal cells and feeding channels on *Nymphaea odorata* leaf. Photographs by David E. Maschwitz, Minneapolis, Minnesota.

Distributor
Ohio Biological Survey, 1315 Kinnear Road, Columbus, Ohio 43212-1192 U.S.A.

Copyright
© 2000 by the Ohio Biological Survey

All rights reserved. No part of this publication may be reproduced, stored in a computerized system, or published in any form or in any manner, including electronic, mechanical, reprographic, or photographic, without prior written permission from the publishers, Ohio Biological Survey, 1315 Kinnear Road, Columbus, Ohio 43212-1192 U.S.A.

Layout and Design: Brian J. Armitage, Ohio Biological Survey
Printing: The Ohio State University, Printing Services, Columbus, Ohio

Ohio Biological Survey
College of Biological Sciences
The Ohio State University
Columbus, Ohio 43212-1192
http://www-obs.biosci.ohio-state.edu

04-2000—0.75 M

Abstract

The subgenera *Polypedilum* (*Polypedilum*) Kieffer and *P.* (*Uresipedilum*) Oyewo and Sæther (Diptera: Chironomidae) are revised for the Nearctic region. Four subgenera in the genus *Polypedilum*, *P.* (*Polypedilum*) Kieffer, *P.* (*Pentapedilum*) Kieffer, *P.* (*Tripodura*) Townes, and *P.* (*Uresipedilum*) Oyewo and Sæther are recognized. Adult males for 30 (*P.* (*P.*) *calopterum* is not keyed nor described), larvae for 15, and pupae for 17, of the 31 species in the two subgenera are described. Adult females are not treated. Nine species are described as new: *P.* (*P.*) *angulum* sp. nov., *P.* (*P.*) *atrichon* sp. nov., *P.* (*P.*) *bergi* sp. nov., *P.* (*P.*) *castornama* sp. nov., *P.* (*P.*) *crassum* sp. nov., *P.* (*P.*) *falciforme* sp. nov., *P.* (*P.*) *nymphaeorum* sp. nov., *P.* (*P.*) *prolixipartum* sp. nov., and *P.* (*P.*) *tuberculum* sp. nov. Eighteen species in *P.* (*Polypedilum*) fit into three species groups based mostly on adult characters. Adult, larval, and pupal characters can show significant disagreement in species associations. Keys to species are included for larvae and pupae in the genus *Polypedilum* and to the species of adult males in the subgenera *P.* (*Polypedilum*) and *P.* (*Uresipedilum*). Adult male *Polypedilum* (*sensu lato*) of the Nearctic region can be separated from most other Chironominae genera by the triangular shaped (narrowed anteriorly) eighth abdominal tergite, the strong, evenly spaced, gonostylar medial setae, and the bifid pulvilli. Some larval and pupal *Polypedilum* can only be separated from other Chironominae genera by a combination of characters. The genus is found nearly world wide and some species are very common. Larvae live in a very broad range of aquatic habitats.

Acknowledgements

The senior author completed this work under the guidance of Dr. Edwin F. Cook, University of Minnesota (retired) to whom I am indebted for his help throughout the study. The study was funded by the National Institute of Health (NIH Grant No. AI-07029). I wish to thank Kris K. Kirkeby for labeling the drawings, and Drs. Roger D. Price, James C. Underhill, Donald R. Oliver, Ole A. Sæther, and James E. Sublette, for providing comments on the thesis.

Sincere gratitude is extended to Dr. Brian J. Armitage and the Ohio Biological Survey for making the publication of this work possible, and to Drs. John H. Epler and Ole A. Sæther for reviewing the manuscript and providing invaluable comments. I wish to acknowledge the contributions of Michael. J. Bolton, Ohio Environmental Protection Agency, who put me in touch with Dr. Armitage and sent me copies of recent literature; Dr. Ralph W. Holzenthal, University of Minnesota, for use of his scanner and other kindnesses; and to Linda A. Maschwitz for helping proof the manuscript.

It is my pleasure to dedicate this publication to the late Dr. Jean L. Laffoon (Iowa State University) who nurtured my interest in academics in general and in entomology in particular.

Most of the specimens borrowed for examination came from the following individuals and museums. (Affiliations listed are at the time the loans were made, 1969-1970).

James E. Sublette, Eastern New Mexico University (JES); Donald R. Oliver, Canadian National Collection (CNC); William M. Beck, Jr. and Elizabeth C. Beck, Florida State Board of Health (Beck); Willis W. Wirth, U.S. National Museum (USNM); L.L. Pechuman, Cornell University (Corn); Henry K. Townes, Jr. (HKT), American Entomological Institute; Andrew L. Hamilton and Ole A. Sæther, Freshwater Institute, Fisheries Research Board of Canada (FRB); Selwyn S. Roback, Academy of Natural Sciences of Philadelphia (ANSP); and Ernst J. Fittkau, Max-Planck-Institut für Limnologie (MPIL).

Additional specimens were obtained from the following individuals and institutions: Jean L. Laffoon, Iowa State University (ISU); H.D. Blocker, Kansas State University (KanSU); George W. Byers, University of Kansas (UKan); Charles L. Hogue, Los Angeles County Museum of Natural History (LAMusNH); Lutz J. Bayer and William L. Hilsenhoff, University of Wisconsin (UW); Donald W. Webb, Illinois Natural History Survey (INH); William T. Mason, Jr., U.S. Environmental Protection Agency, Cincinnati, Ohio (EPACin); Richard W. Fullner, U.S. Environmental Protection Agency, Wheeling, West Virginia (EPAWhg); Patrick L. Hudson, North Central Reservoir Investigations (PLH); Thomas E. Moore, University of Michigan (UMich); John F. Lawrence, Harvard University (Harvard); Saul Frommer, University of California, Riverside (UCR); Wilbur R. Inns, University of Missouri (UMo); Paul H. Arnaud, Jr., California Academy of Sciences (CalAS); E.P. Rouse, University of Arkansas (UArk); K.L. Knight, University of Georgia (UGa); Charles A. Triplehorn, The Ohio State University (OhioSU); P. Oman, Oregon State University (OreSU); P.C. Peterson, University of Nebraska (UNebr); D.M. Rees, University of Utah (UUtah); G.E. Scudder, University of British Columbia (UBC); Roland L. Fischer, Michigan State University (MichSU); F.W. Mead, Florida Department of Agriculture, Division of Plant Industry (FDAgric); George E. Wallace, Carnegie Museum (CMus); R.O. Schuster, University of California, Davis (UCD); M.T. James, Washington State University (WashSU); Pedro W. Wygodzinsky, American Museum of Natural History (AMNH).

Preface

The basis for this work is a Ph.D. thesis completed by the senior author in 1975. The discussion of the literature reflects, to some extent, the state of chironomid knowledge at that time. However, most of the pertinent works published since 1975 involving the genus *Polypedilum* Kieffer are included, except that we have not seen the treatment of Rumanian Chironominae by Albu (1980) and many of the publications by Sasa and his coworkers on Japanese chironomids. Subsequent to the completion of this work, several important changes to the genus have been published. Notably, a new subgenus, *P.* (*Uresipedilum*), was added, and the subgenus *P.* (*Pentapedilum*) was elevated to generic status by Sasa (Sasa and Kikuchi, 1995). Sæther and his coworkers have erected a new *Polypedilum* subgenus, bringing the total subgenera that they recognize to six. Also, two species in the two treated subgenera, *P.*(*U.*) *obtusum* Townes and *P.* (*P.*) *walleyi* Townes, were synonymized by Boesel (1985). The new subgenus, *P.* (*Uresipedilum*), has gained acceptance and it is recognized in this work. However, because we have not reexamined specimens for this publication, we have not recognized nor offered an opinion on the validity of generic status for *P.* (*Pentapedilum*) or the Boesel synonyms. The larval and pupal stages of four species known to us only as adult males have been described, or keyed and figured in post-1975 publications. The four species are *P. pedatum* Townes, *P. albicorne* (Meigen), *P. cultellatum* Goetghebuer (pupal stage known to us), and *P. pedestre* (Meigen) (=*P. fuscipenne* (Meigen)?).

The new species described in the publication should be attributed to the senior author only.

Authors' Note

Saether and Sundal (1999) have designated a new subgenus, *Polypedilum* (*Cerobregma*), that includes *P. ontario* and 16 related species world-wide. *Polypedilum ontario* is the only known Nearctic species in the new subgenus. Because of the many unique characteristics of *P. ontario* adults, larvae and pupae, we agree that subgeneric status for this group of species seems justified. However, we have not included the new subgenus in this work because the publication of Saether and Sundal (1999) appeared too late to be incorporated into this book.

Contents

Introduction and Scope

The genus *Polypedilum* Kieffer comprises a very common group of midges (Diptera: Chironomidae) nearly worldwide in their distribution. Species of *Polypedilum* are found in all zoogeographic regions except the Antarctic. The genus has at least 100 Holarctic species (Cranston *et al.*, 1989). There are many tropical representatives and *Polypedilum* is one of the dominant genera in a rich neotropical chironomid fauna. The genus is uncommon in the Arctic and high mountainous areas.

Adult males of Nearctic *Polypedilum* form an easily recognizable group of species, distinct from males of most other Chironomini genera (Freeman, 1958: 267; Cranston *et al.*, 1989: 405). The males of the Nearctic species of *Polypedilum* can be distinguished by the triangular shaped eighth abdominal segment and the row of strong gonostylar medial setae. Adults of both sexes are recognized ultimately by the deeply bifid pulvilli, a character that can be difficult to see. The genus *Collartomyia* is close to *Polypedilum*; species of both genera have the triangular shaped eighth tergite, but the former can be distinguished from *Polypedilum* and other chironomids by their conelike anteriorly projecting scutum (Amakye and Sæther, 1992; Amakye, 1995).

Four subgenera are recognized in the genus *Polypedilum* Kieffer in the Nearctic region, *P.* (*Polypedilum*) Kieffer, *P.* (*Pentapedilum*) Kieffer, *P.* (*Tripodura*) Townes, and *P.* (*Uresipedilum*) Oyewo and Sæther. This revision covers the Nearctic species in two subgenera, *P.* (*Polypedilum*) Kieffer and *P.* (*Uresipedilum*) Oyewo and Sæther. Thirty-one species in these subgenera are recognized; adult males for 30 and the immature stages for about half of the species are described. *P. (P.) calopterum* is not described. Adult females are not treated. Nine species are described as new. Keys to the species (except *P. calopterum*) of adult males, larvae, and pupae are included. Species in the subgenera *P.* (*Tripodura*) and *P.* (*Pentapedilum*) are not treated, except the immature stages are included in the keys to larvae and pupae. European specimens for the known Holarctic species were examined; however, a more complete examination of the European species might reveal additional Nearctic synonyms. The terminology of the adult, larval, and pupal taxonomic characters follows Sæther (1980). The research for this work was completed in 1975.

The taxonomy of *Polypedilum* adults and immatures has received a fair amount of attention. Johannsen (1905) and Malloch (1915) were the first authors to deal extensively with the Nearctic chironomids and both included the known immatures. Johannsen (1937) and Roback (1957) treated the immature stages for North American species. Townes' (1945) revision of the adult Nearctic Chironomini advanced the knowledge of *Polypedilum* taxonomy substantially.

More recently, an important series edited by Wiederholm (Pinder and Reiss, 1983; 1986; Cranston *et al.*, 1989) provided generic level keys, diagnoses, and taxonomic comments for Holarctic larvae, pupae, and adult males. Bjørio *et al.* (1997), Oyewo and Saether (1998), and Sæther and Sundal (1999) discuss the phylogenetic relationships of *Polypedilum* subgenera and closely related genera. Boesel (1985) reviewed the species of *Polypedilum* known to occur in Ohio and the northeastern part of the United States. Boesel provided keys to adult males and females, larvae and pupae, and discussed the status of most of the species in the genus (see the "Remarks" section in the individual species descriptions). Also, in his review, Boesel (1985) synonymized six species in the genus *Polypedilum*, one in each of the subgenera *P.* (*Polypedilum*) and *P.* (*Uresipedilum*), the remaining four in *P.* (*Tripodura*). These synonymies have not been widely accepted by American authors (*e.g.*, Caldwell *et al.*, 1997; Soponis and Simpson, 1992). Soponis and Russell (1982) and Soponis and Simpson (1992) provided descriptions of adults and immatures, and biological information on several species of *Polypedilum*. Hudson *et al.* (1990) published a list, updated by Caldwell *et al.* (1997), of chironomid species for southeastern United States including notes on the taxonomy and biology of some species of *Polypedilum*. Epler (1995) provided a key to the larval chironomids of Florida including *Polypedilum* with notes on the Florida species. Oliver *et al.* (1990) and Oliver and Dillon (1994) cataloged the Nearctic, and Ashe and Cranston (1991) cataloged the Palaearctic, species of Chironomidae including the *Polypedilum* species.

Adults of the European species have been dealt with by Goetghebuer (1921; 1928; 1937; 1938), Edwards (1929), Brundin (1947), Hirvenoja (1962), Albu (1980) [reference not seen], and Rossaro (1985). Hirvenoja (1962) reviewed several Finnish species in the subgenus *P.* (*Tripodura*). Rossaro (1985) provided quantitative descriptions and keys with figures (except for *P. bicrenatum*) to larvae, pupae, and adults for 14 species known from Italy, seven of which are probably Holarctic. Immature stages of European species have been treated by Lenz (1923; 1941; 1962), Chernovskii (1949), Moller Pillot (1978-1979), and Langton (1984; 1991). Moller Pillot (1978-1979) provided keys to the larval chironomids of the Netherlands, including eight species or species groups of *Polypedilum* with notes on several species. Langton (1984) published a key to the pupal exuviae of British Chironomidae including the known species of *Polypedilum* (15 species plus 7 unnamed taxa). Langton (1991) [not seen] expanded and amended his keys to include western Palaearctic species (19 species, plus 9 unnamed taxa in *Polypedilum*).

Freeman (1958; 1959; 1961) reviewed the African, New Zealand, and Australian species respectively, and Tokunaga (1964) reviewed the Micronesian species. Cranston (1989) and Cranston and Judd (1989) described several new species of *Polypedilum* and documented others from the Middle East and northern Africa. Bidawid and Fittkau (1995) and Bidawid-Kafka (1996) described 51 new species of *Polypedilum* from the Amazon region. Oyewo and Sæther (1998) revise the Afrotropical species in the subgenus *P.* (*Uresipedilum*) and describe ten new species from Africa. Sublette and Sasa (1994) examined chironomids from Guatemala and described 34 new species, including six in the genus *Polypedilum*. Harrison (1996) described the chironomids of Ethiopia. He included nine species of *Polypedilum* (one new species), and identified six larval types within the genus. Sasa has worked extensively with the chironomids of Japan; the authors have not seen most of his publications.

The immature stages of *Polypedilum* are found in an extremely varied and broad range of aquatic habitats. In the north temperate zone, *Polypedilum* larvae are found in brackish and fresh water, in lakes ranging from oligotrophic to eutrophic, wetlands, and in rivers of all sizes. Brundin (1949) and Thienemann (1954) summarized the known biological information for the genus. Some biological information is included in many of the taxonomic papers cited above.

Lake-inhabiting species are most often found in the littoral zone, usually associated with aquatic vegetation. Bottom- dwelling species are usually found in shallow water but some can be found in sediments at considerable depth. A cattail wetland with a variety of floating and submerged aquatic vegetation located near St. Paul, Minnesota (45°05'N 93°08'W) produced six species of *P.* (*Polypedilum*), including three new species.

Stream-dwelling *Polypedilum* are found in bottom substrates, or associated with plants or detritus along the stream edge. Some species live on the surface, or under the bark, of submerged wood in both lakes and streams. Two species live in the dead leaves, algae, and other debris trapped by emergent vegetation or by overhanging branches at the water surface; a habitat which apparently has been further exploited by members of the genus in the Amazon region (Fittkau, 1971: 413). *Polypedilum* are most common in eutrophic waters and several species are described as being facultative in their ability to withstand polluted conditions. *Polypedilum fallax* and *P. flavum* (*P. convictum*) are reported to be pollution tolerant (Paine and Gaufin, 1956; Jankovic, 1973). Leathers (1922b) described the biology of *P. braseniae* in detail, and Berg (1950) described three species associated with pond weeds (*Potamogeton* spp.).

The adaptability of *Polypedilum* larvae is exemplified by the larvae of three species known to be able to tolerate dry conditions and resume normal activities when wetted. The larvae of an African species, *P. vanderplanki* Hinton, live in small water-filled depressions on large rocks. If a depression becomes dry, larvae are able to survive complete desiccation encased in the dry mud at the bottom of the depression. Larvae resume their development when wetted by the next rainfall (Hinton, 1951; 1960). Miller (1969) described another *Polypedilum* species from Nigeria (possibly *P. dewulfi* Goetghebuer) as having the same capabilities of surviving desiccation. *Polypedilum pedatum excelsius* live in a similar, if somewhat less extreme, habitat in central California (Grodhaus and Rotramel, 1980). They found viable eggs and larvae of this species in moist to dry leaf litter in tree-holes.

Swarming of adults has been noted in the literature, but not studied in detail (Harnisch, 1922: 142 Hindes, 1929: 179; Lenz, 1941: 19). Emergence times and patterns have been noted for several species. Harper and Cloutier (1979) described the biology and seasonal emergence patterns of several species of *Polypedilum* living in a system of Laurentian streams in Quebec. *Polypedilum* was the dominant genus of Chironomini in these low pH, low hardness, and bog-stained streams. Soponis (1983) describes the seasonal emergence patterns of five species of *Polypedilum* in a northern Florida stream.

Methods

Source of Specimens

Material for the revision was obtained through specimens loans from the major North American museums (see Table 1), from eight to ten New Jersey mosquito light traps operated weekly throughout Minnesota during the summers of 1968 through 1971, and by collecting adult and immature midges.

Immature stages were associated with adults by rearing individual larvae to the adult stage. Larvae were collected from a variety of aquatic habitats during the summer months, placed in vials with water from the site, one larva per vial, and transported to the lab for rearing. The vials were plugged with a cotton ball, placed in a water bath to provide cooling, and checked periodically. The emerged adult, larval, and pupal exuviae were removed to be mounted on a slide. Larvae were collected for rearing over three summers, 1968-1970. About 17 percent of the larvae collected successfully reached the adult stage, males and females.

Some adults were obtained by "mass rearing" larvae. Plant material, pieces of submerged wood, or other materials thought to harbor *Polypedilum* larvae were brought to the lab and placed in gallon jars containing site water. The jars were fitted with screen tops and aerated to prevent anaerobic conditions. Many adult *Polypedilum* were obtained by this method, but larval and pupal exuviae were not recovered.

Adult males were mounted on slides in balsam with body parts arranged under four cover slips according to the method outlined by Hansen and Cook (1976). Adult males and associated larval and pupal exuviae were mounted on the same slide using six cover slips. All stages were prepared for mounting the same way except the adults were cleared in ten percent sodium hydroxide. Pupal exuviae were always mounted dorsal side up and the larval head capsule ventral side up. It was necessary to crush the head capsule to some extent to expose the mouth parts better. In some cases the frontoclypeal apotome with labrum attached was dissected out and mounted apart from the rest of the head, ventral side up.

Species Descriptions and Figures

All measurements, descriptions, and figures of the three stages are based on slide-mounted specimens. In the species descriptions all measurements are expressed in microns except where indicated otherwise. Precision of measurements for smaller structures, examined at 450X, are within plus or minus three microns. Larval antennal segments, usually measured at l000X, are precise to plus or minus one micron. The first number in a series of values is the mean, followed by the range in parentheses. A second number in parentheses indicates the number of specimens examined for that character if it varied from the number of specimens examined as stated initially. Where reference is not made to length (L), width (W), or diameter (D), the numbers indicate the quantity of the structures being discussed. For example, Dorsocentral S 13.8 (10-21) (6), means the mean number of dorsocentral setae on six specimens is 13.8 with a range of 10-21. The abbreviations used in the descriptions are defined in Table 2 on page 5.

All adult descriptions are of males. The larval and pupal specimens examined included both males and females; the number of each is stated in the larval and pupal descriptions. Adult color descriptions are based on slide-mounted specimens and are not intended to represent the color of live midges. Color is expressed in relative shades of brown, from dark brown to stramineous. The latter term indicates only the very light brown to yellow color of weakly to moderately sclerotized integument. In life these parts are usually white, yellow, or light green. On the other hand, the brown to dark brown color of slide-mounted specimens usually shows through as the same color in life.

The coloration of the pupal exuviae is rather constant within the genus; thus, only variations from the generic description of pupal color are mentioned in the species descriptions. Larval body characters are generally not used because they are difficult or impossible to see on slide mounted exuviae.

The immature stages of *Polypedilum* lack a single character which will separate them from other genera. Thus, it is impossible to write a simple generic diagnosis for *Polypedilum* larvae and pupae. Only a generic diagnosis with exceptions is given for the larvae because they are described in the morphology section.

Table 1. Abbrevations of Institutions Loaning Specimens.

AMNH	American Museum of Natural History
ANSP	Academy of Natural Sciences of Philadelphia
Beck	Florida State Board of Health
CalAS	California Academy of Sciences
CMus	Carnegie Museum
CNC	Canadian National Collection
Corn	Cornell University
EPACin	U.S. Environmental Protection Agency, Cincinnati, Ohio
EPAWhg	U.S. Environmental Protection Agency, Wheeling, West Virginia
FDAgric	Florida Department of Agriculture, Division of Plant Industry
FRB	Freshwater Institute, Fisheries Research Board of Canada
Harvard	Harvard University
HKT	H.K. Townes, Jr. (American Entomological Institute)
INH	Illinois Natural History Survey
ISU	Iowa State University
JES	Eastern New Mexico University (J.E. Sublette)
KanSU	Kansas State University
LAMusNH	Los Angeles County Museum of Natural History
MichSU	Michigan State University
MPIL	Max-Planck-Institut für Limnologie (E.J. Fittkau)
OreSU	Oregon State University
OhioSU	The Ohio State University
PLH	P.L. Hudson, North Central Reservoir Investigations
UArk	University of Arkansas
UBC	University of British Columbia
UCD	University of California, Davis
UCR	University of California, Riverside
UGa	University of Georgia (K.L. Knight)
UKan	University of Kansas
UMich	University of Michigan
UMn	University of Minnesota
UMo	University of Missouri
UNebr	University of Nebraska
USNM	U.S. National Museum
UUtah	University of Utah
UW	University of Wisconsin
WashSU	Washington State University

Table 2. Morphological Characters and Terms Used in Descriptions.

AR	Antennal ratio, adult (Figure 2): length of flagellomere 13 over length of flagellomeres one through 12; larva: length of antennal segments two through five over length of segment one.
BV	Lengths of femur plus tibia plus basitarsus over combined lengths of tarsal segments two through five (Figure 4).
D	Diameter.
Fe	Femur (a)
GsR	Gonostylar ratio: length of gonostylus over width of gonostylus (Figure 5).
HIR	Head width to interocular distance ratio:; head width over interocular distance (Figure 1).
HR	Hypopygium ratio: length of gonostylus over length of posterior lobe gonocoxite (Figure 5).
HV	Hypopygium value: total length of adult male over ten times length of gonostylus (Figure 5).
IVo	Inferior volsella (ae) (Figure 9).
L	Length (*e.g.*, of larval head capsule, Figure 10)
LL	Labral lamella (ae) (Figure 11).
LR	Leg ratio: length of basitarsus over length of tibia (Figure 4).
L total	Total length of adult male; the length of the thorax, as is measured from the antepronotum to the posterior end of the postnotum, plus the length of the abdomen, as measured from the anterior margin of tergite I to the posterior end of the gonostylus.
M	Mentum (Figure 11).
MaxL	Maximum length
PI, PII, PIII	Fore, mid, and hind leg, respectively.
PE	Pecten epipharyngis (Figure 11).
PMa	Pecten mandibularis (Figure 15).
PS	Peg sensillum, sensory seta on antennal segment two of larva (Figure 14).
ROR	Ring organ ratio: length of larval antennal segment one over distance between base of segment one and ring organ Figure 14).
S	Seta (ae)
SM1-4	Maxillary seta (ae) (Figure 16).
SV	Lengths of femur plus tibia over length of basitarsus (Figure 4).
SVo	Superior volsella (ae) (Figure 9).
SVoR	Length of SVo distal to distal seta origin over length of SVo proximal to distal seta origin (Figure 5).
Ta_1	Tarsal segment one (Ta_2=tarsal segment two; Ta_3=tarsal segment three; Ta_4=tarsal segment four; Ta_5=tarsal segment 5)
Ti	Tibia (ae)
VmP	Ventromental plate (Figure 11).
VmPR	Ventromental plate ratio: width of ventromental plate over its length (Figure 10).
VmPSR	Ventromental plate separation ratio: mean width of the two ventromental plates over distance between ventromental plates (Figure 10).
VR	Venarum ratio: length of Cu over length of M (Figure 3).
W	Width

Polypedilum (*P.*) *nubeculosum* (Meigen) is described in the most detail as a representative of the subgenus *P.* (*Polypedilum*). Also, *P.* (*P.*) *illinoense*, *P.* (*P.*) *fallax*, and *P.* (*U.*) *flavum* are described in more detail as representatives of the *illinoense* and *fallax* species groups and the subgenus *P.* (*Uresipedilum*), respectively. These descriptions are often referred to in the descriptions of closely related species.

Most figures of the hypopygia are split into two halves, both dorsal side up. The left half shows surface characters including setae. Hidden characters are shown in dotted line. The right half shows the internal apodemes and sclerotized parts, the ventral basal plate, and the origins of the basal plate medial setae. Figures of the larval mentum show the whole ventromental plate on the left side with visible striations and a partial outline of the right ventromental plate. Figures of the pupal abdomen show tergites II through VIII; the anal lobes and the anal taeniae, shown in Figure 17, are not included. Figures are not shown to the same scale.

External Morphology

Adult Male

Adult structures used in the descriptions and keys are shown in Figures 1-9 and 18-24. Detailed discussions of adult chironomid morphology can be found in Hirvenoja (1973), Hansen and Cook (1976), and Sæther (1971; 1980).

Terminology of the head, thorax, legs, and wing venation follows Hansen and Cook (1976) and Sæther (1971; 1980). However, the wing vein Hansen and Cook (1976) call the vannal fold and Pinder (1989), Oliver and Dillon (1989), and Sæther (1990) call the postcubitus, we call Cu_2 after Bidawid and Fittkau (1995). Also, we call the setae on the

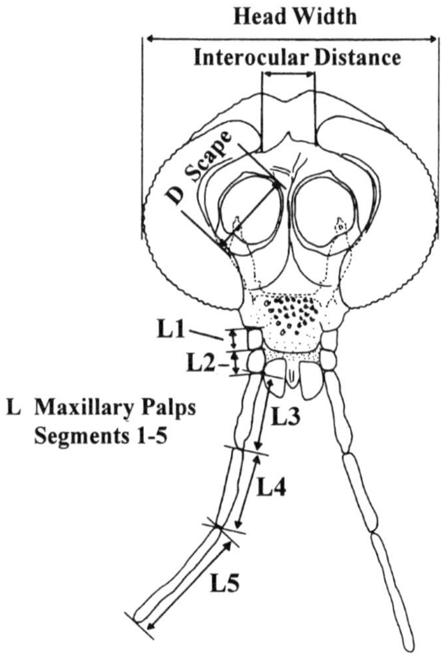

Figure 1. *P. nubeculosum* (Meigen), head showing measurements [**L1-L5** = lengths of maxillary palp segments 1 through 5; **D Scape** = diameter of antennal scape; ratio: **HIR** = head width to interocular distance].

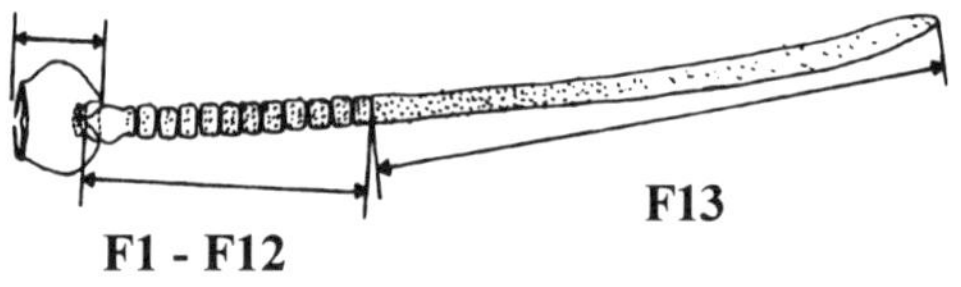

Figure 2. *P. nubeculosum* (Meigen), antenna showing measurements [F1 through F13 = flagellomeres 1-13; ratio: **AR** = length of flagellomere 13 : the combined lengths of flagellomeres 1-12].

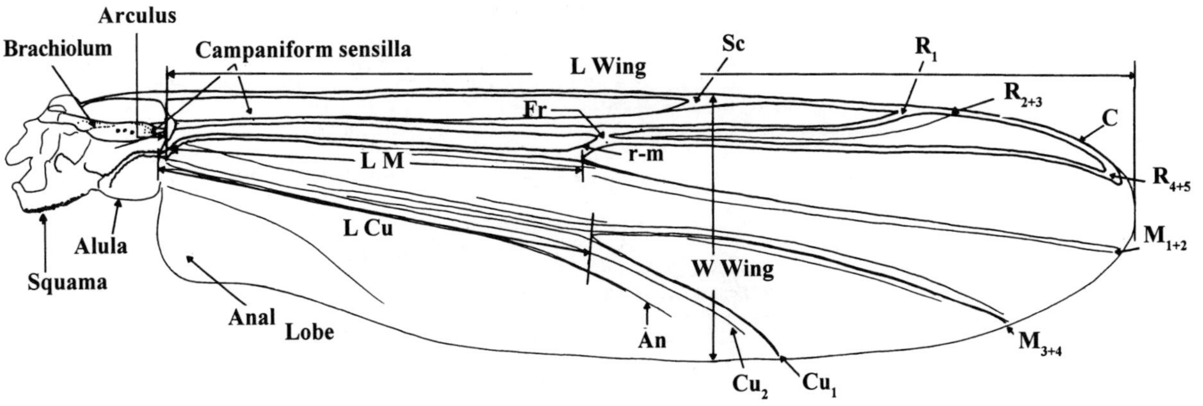

Figure 3. *P. nubeculosum* (Meigen), wing showing measurements [ratio **VR** = length of Cu **:** length of M].

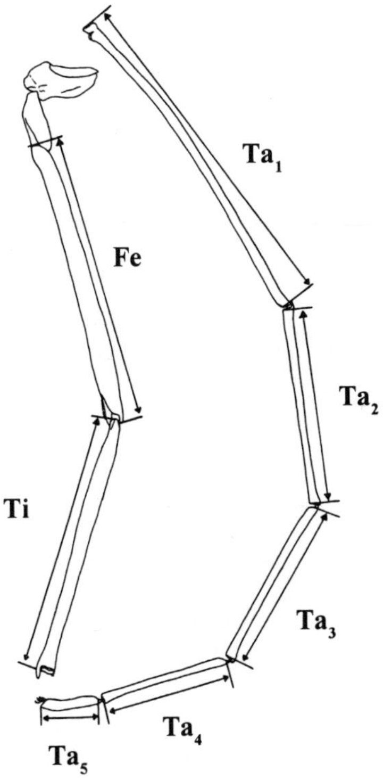

Figure 4. *P. nubeculosum* (Meigen), foreleg, showing measurements. [**Fe** = femur; **Ta_1-Ta_5** = tarsal segments 1 through 5; **Ti** = tibia; ratios: **LR** = leg ratio (length of Ta_1: length of Ti); **BV** = ratio (length of femur, tibia, and tarsal segment 1**:** length of tarsal segments 2-5); **SV** = ratio (length of femur and tibia **:** length of tarsal segment 1)].

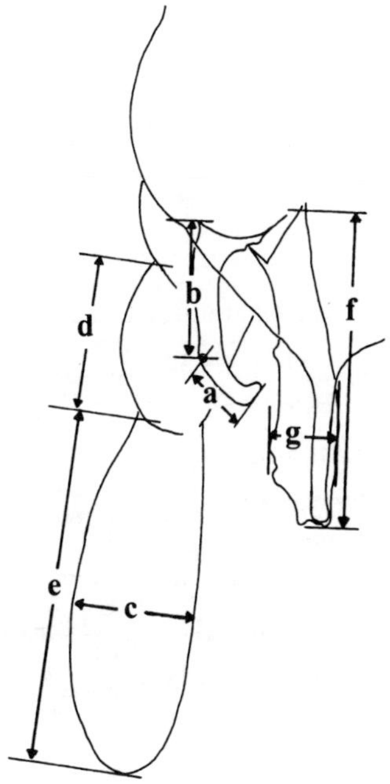

Figure 5. *P. nubeculosum* (Meigen), hypopygium, showing measurements. [**a+b** = length of superior volsella; **c** = width of gonostylus; **d** = length of posterior lobe gonocoxite; **e** = length of gonostylus; **f** = length of inferior volsella; **g** = width of inferior volsella; **SVoR** = superior volsella ratio of **a:b**; **GsR** = gonostylus ratio of **e:c**; and, **HR** = hypopygium ratio of **e:d**].

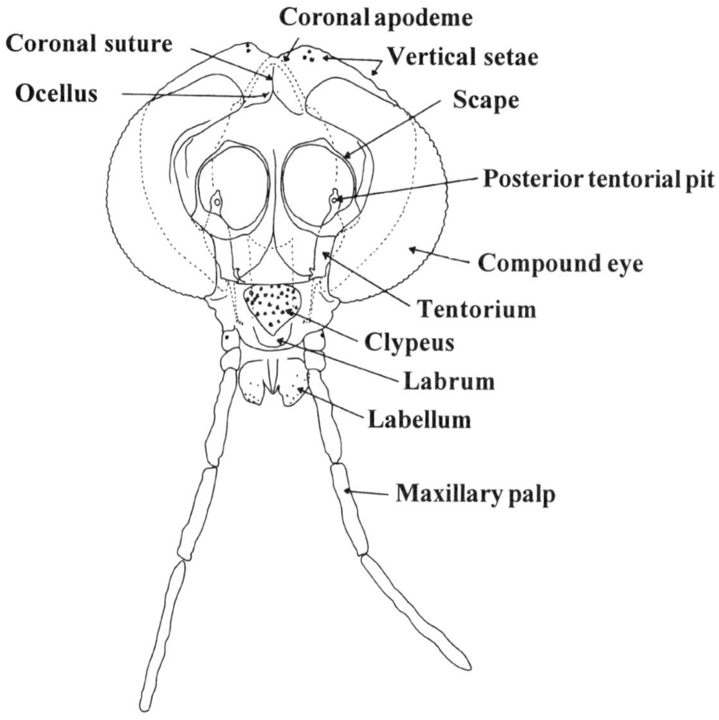

Figure 6. *P. nubeculosum* (Meigen), morphology of head.

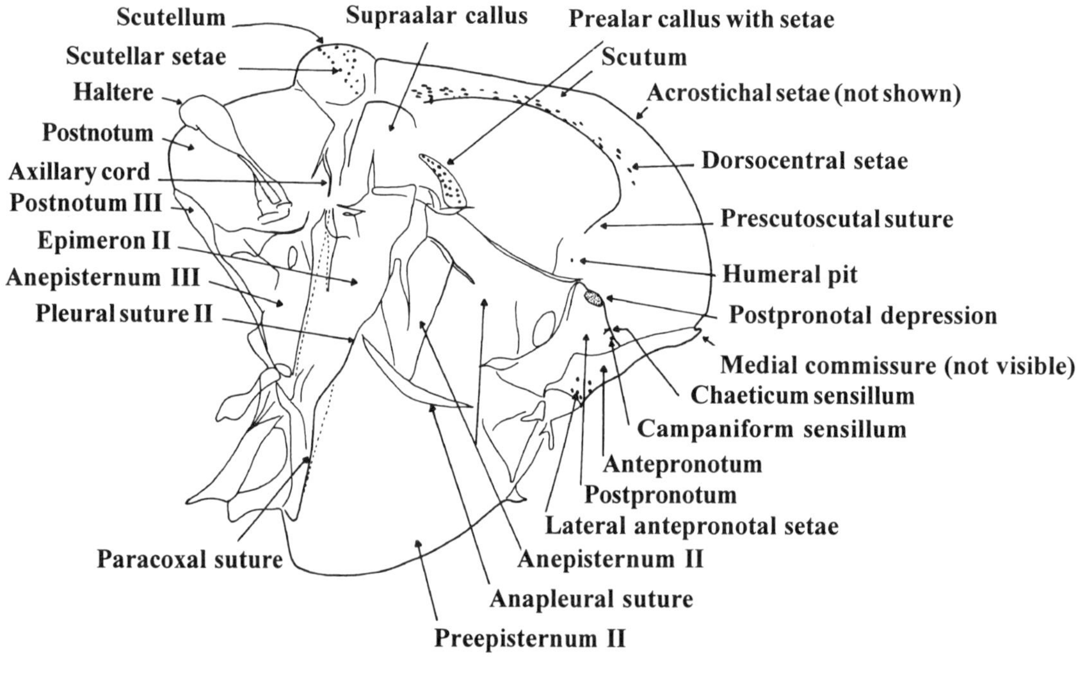

Figure 7. *P. nubeculosum* (Meigen), morphology of thorax.

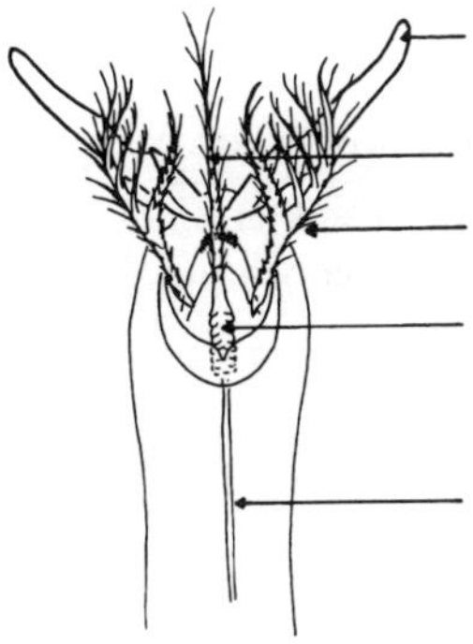

Figure 8. *P. bergi* sp. nov., foretibial apex showing bifid pulvillus,ventral view.

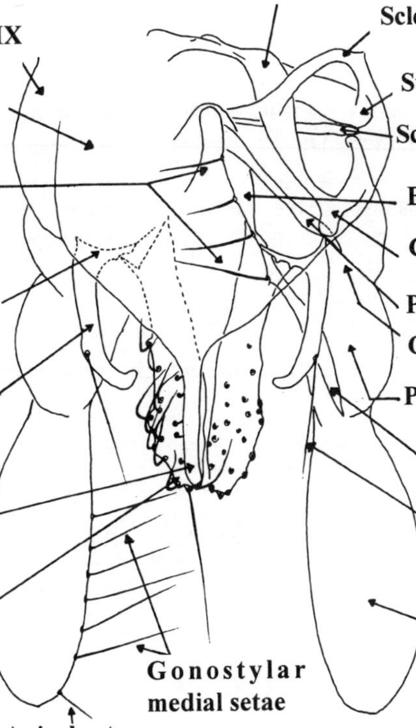

Figure 9. *P. nubeculosum* (Meigen), morphology of hypopygium. (Most hypopygial drawings are split into two halves. The right half is a dorsal surface view showing setae. The left half shows the internal apodemes and sclerotized parts plus the origins of the ventral basal plate, and medial gonostylar setae. Hypopygial drawings not to same scale).

anterolateral part of the thoracic scutum the scutal fossal setae, as do Bidawid and Fittkau (1995). Sæther (1980) and Oliver and Dillon (1989) call these setae the humerals.

The terminology of the male hypopygium follows Sæther (1980). However, we believe that additional morphological work is needed to verify the morphological basis for the current terminology (see *e.g.*, Sæther, 1971; Hirvenoja, 1973; Knight and Laffoon, 1971). Also, homologies between nematocerous groups are not clearly established (van Emden and Hennig, 1970: 136) or are in dispute (Hirvenoja, 1973: 27).

Larva

The terminology for larval structures recommended by Sæther (1980) is widely accepted and is utilized here. Larval structures are shown in Figures 10-16. As in the case of adult hypopygium, we believe that the true homologies of some structures of the larval head capsule are unclear and need further study.

The whole *Polypedilum* larva has been figured and briefly described by Prashad (1918: 71-74, Figure 1) and Lenz (1941: 12, Figure 1, 1962: 237, Figure 352), although from Prashad's figures and description it is impossible to be sure that the larva he described is truly a *Polypedilum*. Living *Polypedilum* larvae range from red or red-orange to greenish yellow in color. Some species, especially those that feed on plant tissue such as *P. braseniae*, show considerable green color. The head is prognathous and usually small in relation to the rest of the body (Lenz, 1941: 12).

Gouin (1957: 109, Figure 1) has enumerated ten setae on the head capsule (Figures 11, 12). *Polypedilum* also have a pair of small setae near the posterodorsal margin of the head, which are not shown in Figures 11 and 12. These setae are the same as seta numbers 11 and 12 on *Cricotopus* (Hirvenoja, 1973: 34, Figure 22). Another pair of setae, the setae submenta (Figure 11), are located just posterior to each ventromental plate (labial seta, Gouin, 1957: 113, Figure 3).

It appears that the dorsal sclerite of the head is not a true morphological sclerite, but a dorsal apotome defined laterally by the "frontal sutures" that are, in reality, the ecdysial cleavage lines of the head (Snodgrass, 1947: 3; 1959: 5). This apotome, at least in the case of *Chironomus*, probably includes all or part of the clypeus as well as the frons based on the fact that seta number 3 is located posterior to the interantennal sulcus (Hirvenoja, 1973: 35). Seta number 3 is similarly located in *Polypedilum*; thus, the interantennal sulcus is assumed to separate the labrum from the frontoclypeal apotome (Figure 12).

Most of the rest of the head capsule is comprised of the genae and post genae (Hirvenoja, 1973: 37). The vertex is reduced by the backward extent of the frontoclypeal apotome and the prognathous orientation of the head. Anteriorly

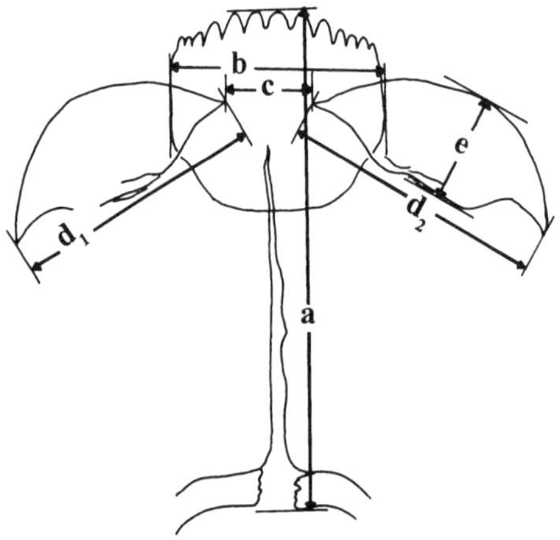

Figure 10. *P. fallax* (Johannsen), larval head capsule showing measurements [**a** = length of head capsule; **b** = width of mentum; **c** = distance between ventromental plates; **d** = width of ventromental plate; **e** = length of ventromental plate; ratios: **VmPR** = **d:e**; **VmPSR** = (average of $\mathbf{d_1}$ and $\mathbf{d_2}$):**c**], ventral view.

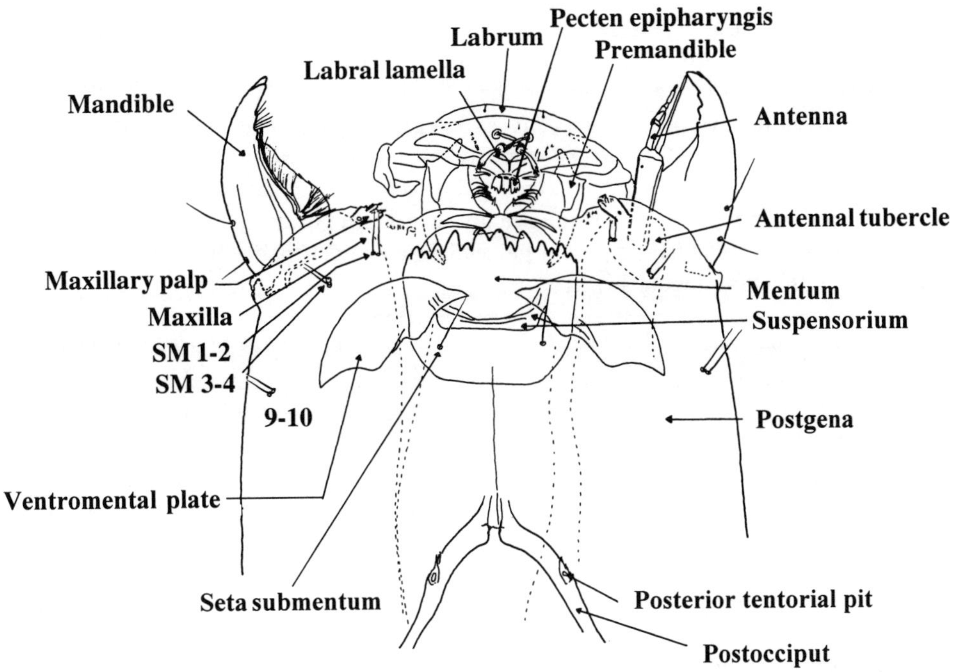

Figure 11. *P. trigonus* Townes, morphology of larval head capsule, ventral view [**9-10** are head capsule setae].

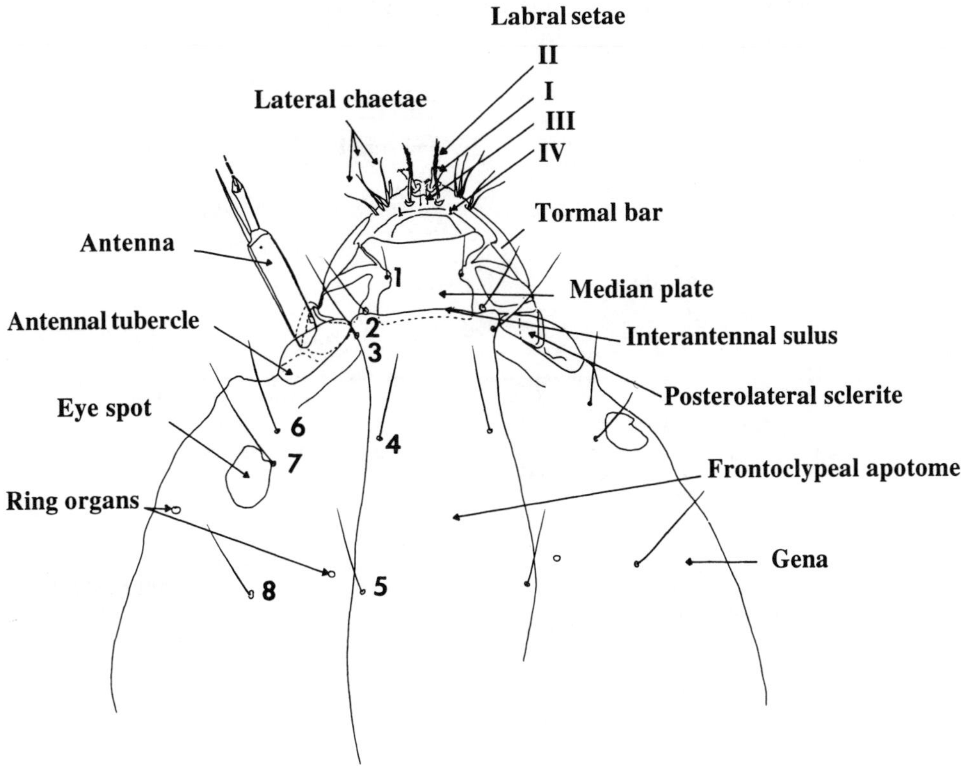

Figure 12. *P. flavum* (Johannsen), morphology of larval labrum, dorsal view [**1-8** are head capsule setae].

the genae bear antennal tubercles, the eye spots, seta numbers 6, 7, and 3, and two ring organs. Lateral to the antennal tubercles, the gena is defined anteriorly by a sclerotized ridge. Theoretically, the pleurostomal suture and the pleurostoma should separate the base of the mandible from the gena (Snodgrass, 1935: 108, Figure 57A). It appears that the pleurostoma may be present in *Glyptotendipes barbipes* (Staeger) (Sublette and Sublette, 1973: Figure 15). The tentorium is also in this area (Hirvenoja, 1973: 36). Anteriorly the postgenae are apparently fused to the cardo of the maxilla, and possibly also to the postmentum of the labium. The ventromental plates are attached to the postgenae in this area. Posteriorly, the postgenae are defined by the heavily sclerotized postocciput; the occiput is apparently lost. The postocciput is set off by the postoccipital suture and surrounds the foramen magnum except dorsally. The posterior tentorial pits are located in this suture (Figure 11).

The morphology of the mentum was reviewed by Sæther (1971), Hirvenoja (1973), and Laffoon and Knight (1973) (the latter for mosquitoes). Sæther (1971) introduced the terms mentum and ventromental plates for the labial plate and paralabial plates, respectively, of most authors prior to 1975. Sæther based these terms on embryological (Craig, 1969) and labial muscle (Cook, 1944b; 1949; Gouin, 1959) evidence that suggests a labial origin for the mentum, plus most of the ventral part of the head capsule as well (Sæther, 1971: 1254). On the other hand, there is support for the theory that the formation of the chironomid prognathous head is not a result of the lengthening of the labium, but a movement and enlargement of parts of the head capsule into the area between the foramen magnum and labium (see Hirvenoja, 1973: 38, 40). According to Snodgrass (1960: 24) the parts of the head involved are the subgenae, the median suture being the result of the two parts meeting medially on the ventral side. But, whether the postmentum (submentum plus mentum) is completely lost as maintained by Snodgrass (1960) and Matsuda (1965: 232), or whether the postmentum is still present, fused to the anterior margin of the postgenae (or the fused hypostomata), is still unclear.

The larval antenna (Figure 14) has five segments. The first bears a ring organ and one or more sense pits, that may be sensilla basiconica (Sublette and Sublette, 1973: Figure 14). The apex of the first segment bears the primary and secondary antennal blades (antennenborste and nebenstift, respectively of Strenzke (1960: 15, 17). The term "blade" is consistent with the fact that these structures are not true setae. The digitiform sensory seta on the apex or subapex of the second segment is called the peg sensillum (Sinnesborste of Strenzke, 1960: 17).

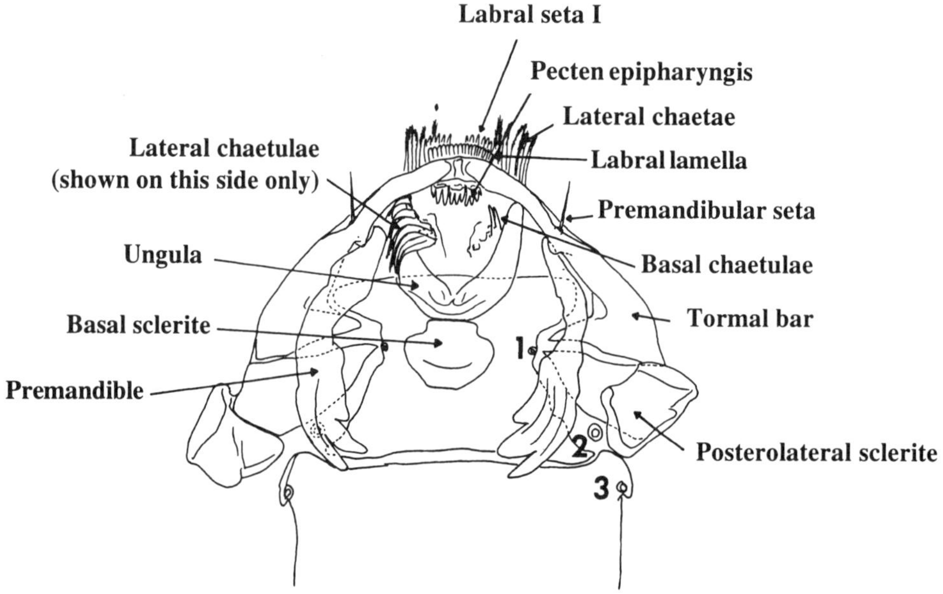

Figure 13. *P. fallax* (Johannsen), morphology of larval labrum, ventral view [**1-3** are head capsule setae].

The labrum in *Polypedilum* is comprised of seven plates or sclerites, including the tormal bars and premandibles (Figures 12, 13). The median plate bears seta numbers 1 and 2 (this is shown as two plates, labral sclerites 1 and 2 in Cranston and Reiss, 1983: Figure 2.2B). The posterolateral sclerites of the labrum (tormae of Cook, 1944a) curve ventrally into the epipharyngeal area (labral sclerites 3, 4, and 5 in Cranston and Reiss, 1983: Figure 2.2B). In *Polypedilum* the posterolateral sclerite usually appears as a single sclerite, but in some specimens, it appears to be divided into two sclerites, one dorsal, the other ventral. This separation is often hidden by the position of the sclerites in slide preparations. The tormal bars are two heavily sclerotized sclerites, anterolateral in position (Sæther, 1980; Cranston and Reiss, 1983). These are the anterolateral sclerites of Goetghebuer (1912: 8) and Bryce (1960: 22), the tormae of Mozley (1970), and the margo labralis of Zavřel (1941: 105). The premandibles articulate on the tormal bars.

Zavřel (1941: 107, Figure 1) provides the basis for much of the terminology of labral and epipharyngeal structures. The anterior membranous field (Bryce, 1960: 23) bears the anterior labral setae I-IV, plus a number of chaetae and spinules that are not true setae. The labral lamella (pecten labralis of Reiss, 1969: 436; the median pectinate chaeta of Gouin, 1957; 113; the chaeta media of Zavřel, 1941: 107) is a large comb-like chaeta anterior to labral seta I. Gouin (1957: 11) and Strenzke (1960: 17) indicate this is a double chaeta in *Chironomus* and *Clunio*, respectively. In *Polypedilum* the origin or base of this chaeta is difficult to see on most slides, but it appears to have a single origin close to, or associated with, setae I, from which it expands anteriorly into the pectinate lateral extensions. One specimen of *P. illinoense* that provided a relatively clear view of the labral lamella had 13 teeth on each side for a total of 26 teeth. Other epipharyngeal structures include the lateral chaetae and spines, the pecten epipharyngis that is divided into three parts, and the ungula that enclose the lateral chaetulae and basal chaetulae. In *Polypedilum* there appears to be six lateral chaetulae on each side. The basal chaetulae, that are usually hidden by the lateral chaetulae, appear as a closely associated pair of chaetulae on each side in such species as *P. fallax* (Figure 13), but they appear more like a deeply bifid single chaetula in *P. flavum*. Located just posterior to the ungula is the basal sclerite.

Questions persist, we believe, of whether the premandibles (Figure 13) represent the true appendages of the premandibular segment or are derived from the labrum, and whether they are homologous with similar structures in

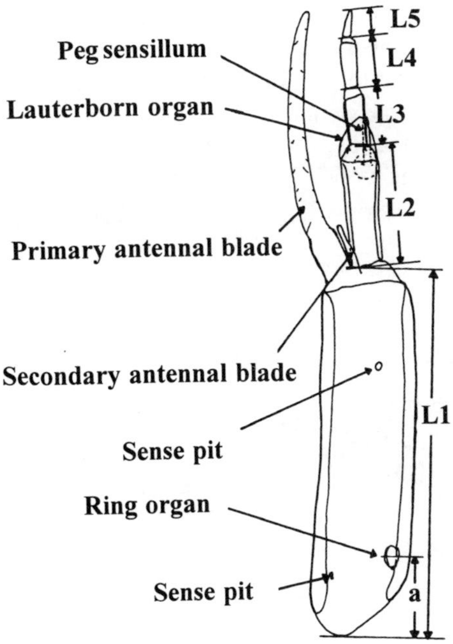

Figure 14. *P. illinoense* (Malloch), measurements and morphology of larval antenna [**a** = the distance from the ring organ to the base of antennal segment 1; **L1-L5** = length of antennal segments 1 through 5; ratio: **ROR** = ring organ ratio (**L1:a**)].

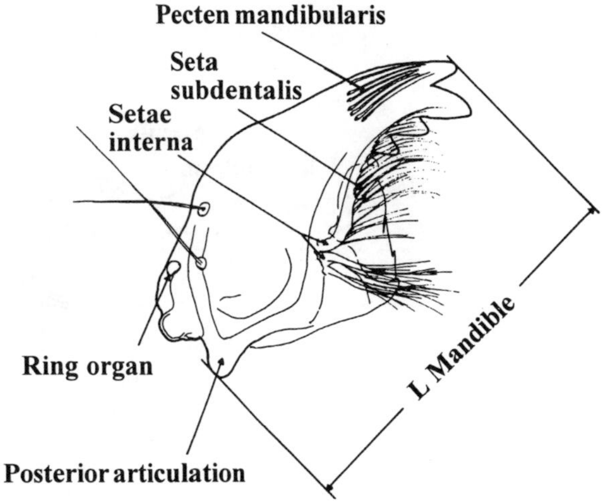

Figure 15. *P. trigonus* Townes, measurements and morphology of larval mandible.

other nematocerous larvae. This was discussed by Sæther (1971) and briefly by Hirvenoja (1973). Both of these authors cite Craig's (1969) paper on Simuliidae in support of the premandibular theory. The term torma has been applied to the premandibles by Anthon (1943), Hennig (1948), Snodgrass (1959), Strenzke (1960), Matsuda (1965), Reiss (1969), Sublette (1964), and Sublette and Sublette (1973). There seems to be some evidence that these structures are derived from the labrum. If that is the case, the question remains whether the premandibles represent the true tormae as indicated especially by Anthon (1943: 18), Hennig (1948: 30), and Snodgrass (1959: 12), or whether they are secondary labral structures, the true tormae being the posterolateral sclerites as indicated by Cook (1944b: 70) (apparently also by Tokunaga, 1932: 10). The posterolateral sclerites are in the conventional tormal position and they bear a single muscle insertion. Tormae are defined by bearing the insertion of the ventral labral muscles (equals the lateral labral retractors?) (Snodgrass, 1935: 113; 1959: 12). These muscles often have two origins and a single insertion (Cook, 1944a: 27). It is possible that these muscle bundles developed separate insertions that migrated apart to form the muscle pairs that are found on the chironomid premandibles and mosquito mouth brushes. Also, it is possible that the dorsal labral muscles have moved laterally to the position on the posterolateral sclerites as suggested by Snodgrass (1959: 13). This would account for the presence of two muscle insertions on the premandibles. Until these uncertainties are resolved, the commonly accepted term, premandible, is used.

The mandibles (Figure 15) have four dark-colored teeth with an apparent fifth tooth fused proximally to the medial margin of the mandible in some species. The mandibles are heavily sclerotized and the posterior (ventral in position) articulation or condyle is very prominent (Zavřel, 1941: 3, Figure 1c, apparently has the posterior and anterior mandibular articulations reversed). The outer margin has two setae and a ring organ. The inner or medial margin is usually serrate and also bears the seta subdentalis and seta interna. The four main branches of the seta interna appear in some species to have more than one base. *Polypedilum nubeculosum* especially has a seta interna (separate setae?) with each branch having its own base or origin, although three of the four are very closely associated.

The main part of the maxilla (Figure 16), or at least the part bearing setae SM 1 and 2 (Gouin, 1957) has been called the stipes (Gouin, 1957: 123; Cook, 1944b: 71) and the palpiger (Hirvenoja, 1973: 39). Because a separate stipes and palpiger cannot be recognized, we prefer the term stipes. In addition, the palpiger seems to be a secondary structure or simply that part of the stipes that bears the palp (Matsuda, 1965: 69). Medially and somewhat dorsally from the palp is a slightly raised area that has been called the galea by a number of authors (Mozley, 1971). Still farther medially is the lacinia. The lacinia has usually been considered a medial lobe bearing a number of chaetae in Chironominae as in the other chironomid subfamilies (Lenz, 1941, 1962; Bryce, 1960; Reiss, 1969; Mozley, 1971; Sublette and Sublette, 1973). It appears that the long medial, often beak-like appearing, "chaetae" of the lacinia in *Polypedilum* are actually deeply bifid extensions of the lacinia. They appear to be part of the lacinia, and they are considered as such here. Gouin (1957: 113, Figure 3) and Leathers (1922b: 53, Figure 11) have figured the lacinia of *Chironomus* and *P. braseniae* in

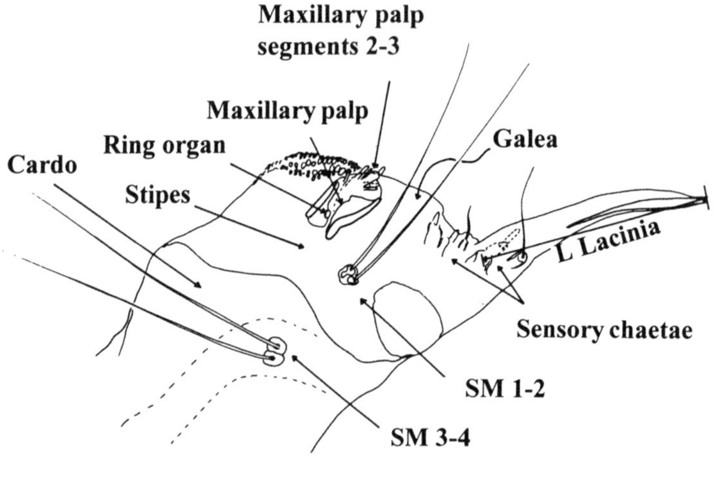

Figure 16. *P. fallax* (Johannsen), measurements and morphology of larval maxilla [**SM1-SM4** = maxillary setae 1 through 4].

this manner, respectively. Lenz (1962: 238, Figure 359) and Kugler and Reiss (1973: 69, Figure 7b) use the term "borsten" for these medial extensions of the lacinia.

The cardo is apparently fused to the hypostoma (subgena), and is indicated by the presence of setae SM 3 and 4. The maxilla and maxillary palp bear an elaborate array of sensilla, setae, and chaetae that have been described and numbered by Mozley (1971). The maxillary palps of *Polypedilum*, as in other genera (see *e.g.*, Sublette and Sublette, 1973: Figure 37), appear to be three segmented (Sublette, 1957a: 38; Kugler and Reiss, 1973: 71). These segments have been interpreted as sensilla basiconica on a large base (Mozley, 1971: 300).

Because the prementum (hypopharnyx) is not used in this study, and because it is often invisible in the typical head capsule preparation, it is not figured. (See Mozley, 1971; Hirvenoja, 1973).

Pupa

The terminology of the pupa used in this work follows Sæther (1980), as illustrated in Coffman (1986: 11). Pupal structures are shown in Figures 17 and 74 elsewhere in the text.

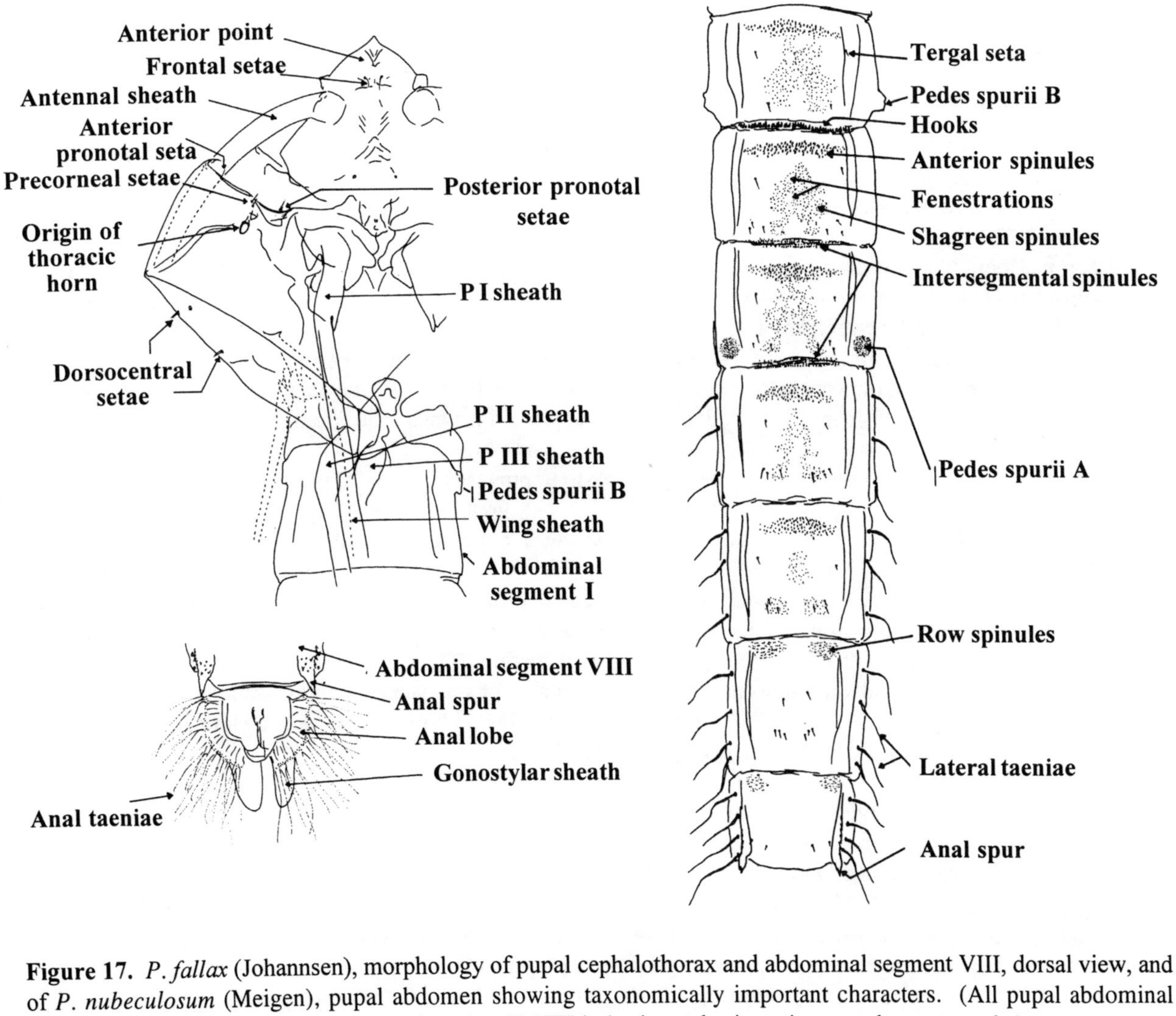

Figure 17. *P. fallax* (Johannsen), morphology of pupal cephalothorax and abdominal segment VIII, dorsal view, and of *P. nubeculosum* (Meigen), pupal abdomen showing taxonomically important characters. (All pupal abdominal drawings that follow are dorsal views of tergites II-VIII in horizontal orientation, not drawn to scale.).

Cephalothorax: The terminology of the cephalothoracic setae is based on Brundin (1966: 88) and Sæther (1971: 1253) (Figure 17). The anterior point (frontal area) has two low swellings bearing the frontal setae (preapical setae of many North American authors). These are the largest setae on the cephalothorax. The swellings that bear these setae are developed into the setose cephalic tubercles in some genera such as *Stictochironomus*. *Polypedilum ontario* has very distinct cone-shaped cephalic tubercles (see Figure 118; page 102). At least one population in the species *P.* (*Tripodura*) *simulans* Townes has well developed cephalic tubercles without setae located anterior to the setate swellings (Figures 31, 32). The base of these cephalic tubercles seems to be the part of the integument that covers the adult pedicel. The pronotal area has an anterior seta and a posterior pair of setae of unequal size. A pair of precorneal setae is located just anterior to the origin of the thoracic horn. An anterior and posterior pair of dorsocentral setae are located generally in the area of the adult dorsocentral setae (Figure 17).

The thoracic horns (respiratory organs) of *Polypedilum* are very thin walled and tube-like. Based on the pre-1975 literature, *Polypedilum* thoracic horns have from four to 12 branches (16 in Pinder and Reiss, 1986: 346). Johannsen (1937: 30) and Roback (1957: 114) both report that the thoracic horns of *Polypedilum* have four to eight branches. Hauber (1947: 464) states that *P. illinoense* has a thoracic horn with six branches. Lenz (1941: 13, 1962: 238) uses the number and arrangement of thoracic horn branches to separate *Polypedilum* into three groups. Those with only four relatively short branches form the *convictum* group (=*P.* (*Uresipedilum*)?); those with eight branches that branch from the origin in a regular arrangement form the *nubeculosum* group; and those with eight-12 branches that vary in length and branch from the origin in an irregular arrangement form the *laetum* group.

Polypedilum bergi sp. nov., *P. falciforme* sp. nov., and *P. cultellatum* Goetghebuer are unique among the *Polypedilum* studied in that they have a distinct prealar tubercle (see Figure 74, page 69).

Abdomen: Zavřel (1942: 30) described the posterolateral lobe or swelling on abdominal segment II (pedes spurii B) and the whorl of spinules on the posterolateral corner of sternite IV (pedes spurii A) as remnants of pseudopodia. Pedes spurii B also occur on segment I of most species of *Polypedilum.*

Abdominal tergites and sternites bear both setae and spines of various sizes, but only the features of the tergites will be considered here (see Figure 17; page 15). Pagast (1947: 504) described two types of spines ("Chagrinierung"), in addition to the dorsal, ventral, and L setae for species of *Diamesa.* Fittkau (1962: 50) has described four types of setae for the Tanypodinae. However, neither of these systems is entirely satisfactory for *Polypedilum.* Abdominal setae include the very small tergal (and sternal) setae, the lateral L setae on segments II-IV, the much larger lateral taeniae on segments V-VIII, and the anal taeniae on the anal lobes (Langton, 1994).

Tergal spines are essentially of two types, the large hook-like spines (called hooks) that always form a uniserial row on the posterior margin of tergite II, and the spinules that vary greatly in size. Tergite I has setae but no spines. Large spinules form the multiserial rows or bands on the anterior margins of tergites II through VI (III-VI in some *P.* (*Uresipedilum*) species). Small to very small spinules form the shagreen pattern in the central parts of tergites II through VI. The points on these spinules are always directed posteriorly. Spinules are usually present in the intersegmental membrane between segments III and IV, and IV and V (Lenz, 1962: 239, Figures 363 and 364). These spinules may appear to be on the extreme posterior margin of tergites III and IV, and they usually point anteriorly. Tergites VII and VIII often have dorsolateral patches of very small spinules arranged in rows of two to four spinules (Lenz 1962: 239, Figure 365). The shagreen spinules are bordered laterally by longitudinal lines that represent the location of muscle attachments (Zavřel, 1942: 31).

On the posterolateral margins of segment VIII are the dark colored anal spurs (absent in *P. braseniae*). *Polypedilum* usually have an anal spur with a single primary point with none to several apical accessory points (see Figures 34 and 35, page 32). The base of the anal spur of some species also bears small lateral accessory points (see Figures 80, 104, and 119). The anal segment has flattened extensions on each side, the anal lobes, which bear a fringe of large lamelliform setae, the anal taeniae. Usually visible on exuviae, extending posteriorly from the ventral side of the anal segment, are two sac-like sheaths (Figure 17). In males the sheaths extend well beyond the tips of the anal lobes and are the coverings of the gonostyli. Similar, but shorter, sheaths in females are the coverings of the cerci.

Measurements, Ratios, and Evaluation of Characters

Adult Male: Adult structures used in the descriptions and keys are shown in Figures 1-9 and 18-24. The methods of taking measurements and determining ratios generally follow Sæther (1969; 1971), Schlee (1966), and Strenzke (1959). The length (L) of the thorax is measured from the antepronotum to the posterior end of the postnotum; the length of the abdomen is measured from the anterior margin of tergite I to the posterior end of the gonostylus. The total length (L total) of the male is the sum of these two measurements.

Figure 1 shows the head measurements. The width (W) of the head is its maximum width in a frontal view; the interocular distance is measured between the medial extensions of the compound eyes. The head width to interocular distance ratio (HIR) is the head width divided by the interocular distance. This character proved to be highly variable intraspecifically and of limited value. Also, it is easily altered by distortions of the head in poorly mounted specimens. Each row of vertical setae on the head was counted as a single unit. No distinction was made between inner and outer vertical setae. In general, the relative lengths of the maxillary palp segments are quite uniform within the two subgenera; however, differences in the lengths of segments three and four have some value as a taxonomic character.

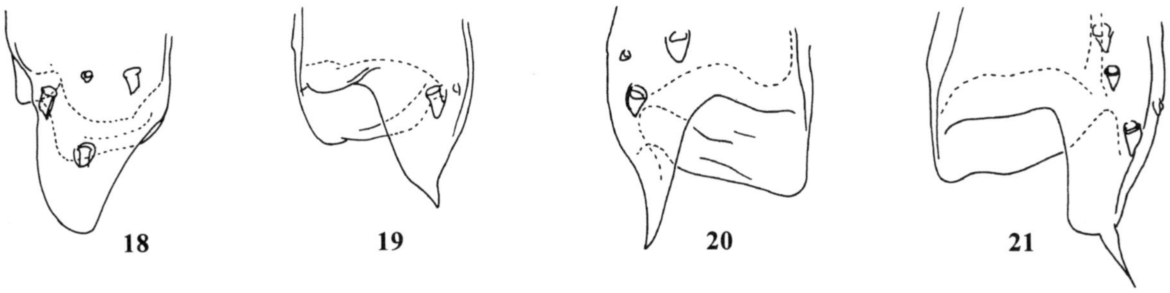

Figures 18-21. 18-*P. crassum* sp. nov., foretibial apical scale; 19-*P. nymphaeorum* n.sp., foretibial apical scale; 20-*P. fallax* (Johannsen), foretibial apical scale; 21-*P. prolixipartum* sp. nov., foretibial apical scale.

The diameter (D) of the scape was measured through its maximum reach (Figure 1), flagellomeres 1-12 were measured from the base of the first flagellomere, which is recessed into the apical depression on the pedicel (Figure 2).

Wing length is measured from the arculus to the apex; wing width is its maximum width and the venarum ratio (VR) is determined as shown in Figure 3. The number of sensilla campaniformis on the brachiolum, Sc, R_1, and near the fork of the radial sector and R_1 (Fr) is quite uniform within the two subgenera. However, the distance between the Fr and R_1 sensilla is highly variable intraspecifically. Brachiolar, R_1 and R_{4+5} setae were counted, but these counts showed considerable intraspecific variation. Those species with wing markings usually show some variation in the darkness or distinctness of the markings.

Parameters used in measuring the legs and leg ratios are shown in Figure 4. The foretibial apical scale with or without its apical spine is a useful character in some species groups (Figures 18-21). The apical combs with their spines on the mid and hind tibia show little variation within the two subgenera.

Measurements of the parts of the hypopygium are shown in Figure 5. The distance between the apex of the anal point and the apices of the inferior volsellae (IVo) is expressed as a positive value if the anal point exceeds the inferior volsellae posteriorly, and as a negative value if the inferior volsellae exceed the anal point. This character shows considerable intraspecific variation, and its measurement is affected by misorientation of the hypopygium on poorly prepared slides. The superior volsella ratio (SVoR) is the portion of this structure beyond the distal seta origin divided by the portion proximal to the distal seta origin. This character is highly variable in some species, indicating variation in the placement of the distal seta, but rather constant in other species. The length of both superior volsellae (SVo) was measured and a mean value obtained. Mean values were also used for the lengths and widths of the two inferior volsellae. Accurate measurements of the inferior volsella width was impossible in some specimens if the hypopygium was tilted on the slide. This measurement proved to have very limited value. The hypopygium ratio (HR = length of gonostylus divided by length of posterior lobe gonocoxite) and hypopygium value (HV = total length divided by 10 times length of gonostylus) are fairly constant within species groups, but somewhat variable intraspecifically. The gonostylus ratio (GsR) is the length of the gonostylus divided by its width. The strong setae on the gonocoxite medial ridge and the very large microtrichia on the gonocoxite basal plate (Figure 9) are quite constant in the two subgenera, and for simplicity, are left off the hypopygial drawings.

The form of the various parts of the hypopygium provides the best characters for distinguishing between *Polypedilum* species (see Strenzke, 1959: 8-10). Separation often is based primarily on differences in the superior volsellae in closely related species (*e.g.*, the *illinoense* group). Differences in the inferior volsellae and gonostyli, plus most of the ratios (AR, LR, SVoR, HR), are often more useful for separating species groups than individual species. The AR and LR vary intraspecifically.

The number of setae (S) on the thorax and hypopygium varies widely within species (see Figures 22-24). Generally, larger specimens have more setae and smaller ones fewer setae. The presence or absence of setal groups such as pronotal setae, scutal fossal setae, and the number of brachiolar setae is of some value. Also, the relative maximum lengths of the distal seta and the setae in a row on the inner margin of the gonostylus (gonostylar medial setae) are fairly good characters.

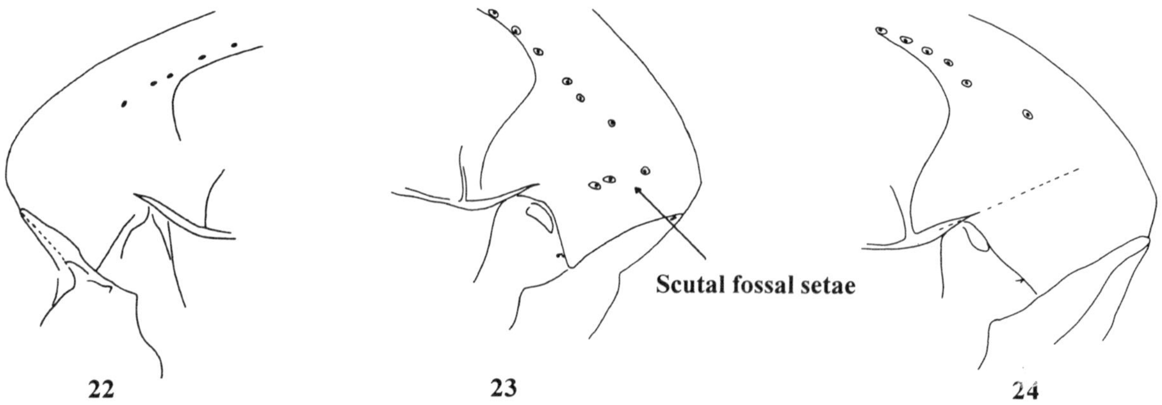

Figures 22-24. 22-*P. illinoense* (Malloch), antepronotum and scutum, lateral view; 23-*P. fallax* (Johannsen), scutum showing scutal fossal setae; 24-*P. prolixipartum* sp. nov., scutum showing dorsocentral setae posterior to line extended through prescutoscutal suture.

Larva: Larval structures used in the descriptions and keys are shown in Figures 10-16. All references are to fourth instar larvae. Parameters used in measuring the head capsule are shown in Figure 10. The head length is the distance from the anterior edge of the mentum (M) to the posterior margin of the head along the mid-ventral line. Because the mentum curves dorsally at its lateral margins, the width of the mentum is significantly increased after the head capsule is flattened on the slide. The measurements recorded are for flattened head capsules, but little value can be attached to this character because of the difficulty in standardizing the degree to which the head has been flattened. The distance between the ventromental plates (VmP) is not significantly affected by flattening of the head capsule. The ventromental plate ratio (VmPR) is the width divided by the length of the ventromental plate. The ventromental plate separation ratio (VmPSR) is the mean width of the two ventromental plates divided by the distance between the ventromental plates (Figure 10). Antennal segments were measured as shown in Figure 14. The location of the ring organ is expressed by the ring organ ratio (ROR) which is the length of the first segment divided by the distance from the base of the first segment to the ring organ.

The length of the mandible is measured from the posterior articulation to the apex (Figure 15). Lacinia length is measured as shown in Figure 16.

The shape of the ventromental plates, VmPR, VmPSR, and lengths of the antennal segments, blades, and Lauterborn organs are the best characters for separating species. The heights of the various teeth of the mentum, development of the posterior lobe of the ventromental plates, and shape of the labral setae usually vary only from one species group to another.

Pupa: Pupal structures used in the descriptions and keys are shown in Figures 17 and 74. The length of the cephalothorax is measured from its anterior apex to its junction with the abdomen; the abdominal length continues from that point to the apex of the anal lobe, not including the anal taeniae. The number of branches on the thoracic horn cannot be determined accurately on slide mounted exuviae; therefore, this character has not been used. It should be noted before the specimen is mounted.

The position and number of abdominal tergal setae seem quite constant within species in the two subgenera. The shagreen spinule pattern on the tergites is often interrupted by small circular bare patches called fenestrations (see Figure 98; page 86). The size and arrangement of fenestrations are useful characters in the separation of some species. One has to be careful not to confuse tergal and sternal setae on slide mounted pupal exuviae.

The variations of the anal spurs on the posterolateral margins of segment VIII have some diagnostic value. However, the number and size of the apical accessory points on the anal spurs are highly variable within most *Polypedilum* species and of little or no diagnostic value, although they have been mentioned as important characters in the past. The number of anal taeniae also varies considerably in most species.

In general, the best pupal characters for species separation are the shagreen patterns on tergites II-VI, the relative sizes of the hooks on tergite II and the tergal spinules, and the size and presence or absence of the intersegmental spinules.

Systematics

Literature

Kieffer (1913: 15, 25) described two new genera, *Polypedilum* and *Pentapedilum*. He designated *P. emarginatum* Kieffer (syn. *P. nubeculosum* (Meigen)) as the type species for *Polypedilum*, and *P. stratiotale* Kieffer (syn. *P. tritum* (Walker)) as the type species for *Pentapedilum* (Kieffer, 1913). However, Ashe (1981) discovered that Kieffer had described the genus *Polypedilum* in a paper published in 1912. In this publication Kieffer put two species in *Polypedilum*, *P. pelostolum* and *P. ceylanicum* (Kieffer, 1912: 41, 42). Ashe (1981) concluded that these names were mistakenly published prior to the naming of the new genus in the 1913 paper because Kieffer had not designated one of these two species as the type for the genus. Because the names in Kieffer's earlier publication have priority, Ashe selected one of the species, *P. pelostolum* Kieffer, as the type species for the genus *Polypedilum*. *Polypedilum pelostolum* is a synonym of *P. nubifer* (Skuse) (Freeman, 1961: 707).

Polypedilum nubifer (Skuse), now the type species for the genus, is considered atypical for the genus in all life stages (Cranston *et al.*,1989). *Polypedilum nubifer* is discussed and figured by Freeman (1961) and Cranston *et al.* (1989) (adult), Pinder and Reiss (1983) (larva), and Pinder and Reiss (1986) (pupa). *Polypedilum nubifer* is a widely distributed Palaearctic and Oriental species (Ashe and Cranston, 1991).

Kieffer (1913) based both new genera, *Polypedilum* and *Pentapedilum*, on the presence of bifid pulvilli. He indicated, however, that *Pentapedilum* was close to *Tanytarsus*, presumably because of the presence of wing setae (macrotrichia). Subsequently, Kieffer named several new genera which are synonyms of either *Polypedilum* or *Pentapedilum* (see Freeman 1958: 266-270). Apparently none of these Kieffer genera contained Nearctic representatives. Freeman had access to most of Kieffer's material and was able to identify most of his species and genera.

The first authors after Kieffer to recognize *Polypedilum* as a separate genus were Goetghebuer (1921) and Lenz (1921). Goetghebuer (1928; 1938) put *Polypedilum* in the tribe Chironomini. Goetghebuer (1928) treated the genus *Pentapedilum* in a very broad sense. He included in it the subgenera *P.* (*Clinotanytarsus*), *P.* (*Sergentia*), *P.* (*Kiefferulus*), *P.* (*Phaenopsectra*), and *P.* (*Pentapedilum*). He placed *Pentapedilum* in the tribe Tanytarsini based solely on the presence of setate wings. In 1938 Goetghebuer raised *Pentapedilum* (*sensu stricto*) from subgeneric to generic level, but retained it in Tanytarsini (Goetghebuer, 1938).

Edwards (1929: 406) noted that *Pentapedilum* (*sensu lato*), which have setae on both the wings and squamae, are much more closely related to *Chironomus* (*sensu lato*) than Tanytarsini. Therefore, he treated *Pentapedilum*, which included the subgenus *P.* (*Pentapedilum*), as one of his five genera of Chironominae, and *Polypedilum* as a subgenus of *Chironomus*. By 1931, Edwards felt that the presence or absence of setae on the wings had been overrated as a character; he stated that "this feature by itself can have at most subgeneric value" (Edwards, 1931: 310). He further noted that *Pentapedilum* (*sensu stricto*) is "essentially similar" to *Polypedilum*, and he restricted the genus *Pentapedilum* (Chironomini) to include only those species in his 1929 subgenus *P.* (*Pentapedilum*) (Edwards, 1931: 311).

Townes (1945) recognized that *Pentapedilum* (*sensu stricto*) species were very close to *Polypedilum* in that adult males shared three important diagnostic characters (bifid pulvilli, triangular shaped eighth tergite, and strong gonostylar medial setae); thus, he designated *Pentapedilum* (*sensu stricto*) a subgenus of *Polypedilum*.

The subgeneric rank for *Pentapedilum* (*sensu stricto*) has been accepted by most subsequent authors. However, initially, Townes' third subgenus *P.* (*Tripodura*) was not. *Polypedilum* (*Tripodura*) was created to include those species with a gap apically between the R_1 and R_{2+3} wing veins, broad pad-like superior volsellae with multiple apical setae, and often lateral points or shoulders on each side of the base of the anal point (Townes, 1945). The separation of R_1 and R_{2+3} is variable within the genus and is probably the poorest of the three characters. Freeman (1958: 269) synonymized *P.* (*Tripodura*) with *P.* (*Polypedilum*) because of "too much variation and inter grading" between the two groups. He further stated that the genus "cannot easily be subdivided into subgenera although quite well marked species groups can be seen."

Most authors (prior to 1975) followed Freeman with regard to *P.* (*Tripodura*) (Dendy and Sublette, 1959; Sublette and Sublette, 1965; Tokunaga, 1964; Fittkau *et al.*, 1967; and Hamilton *et al.*, 1969). Hirvenoja (1962) was the major exception to this point of view. Rossaro (1985) recognized the three subgenera in his revision of the Italian *Polypedilum*. On the other hand, Boesel (1985) concluded that any division of *Polypedilum* into subgenera should be postponed until information on the immature stages is more complete. We feel that the separation between the Nearctic species of *P.* (*Tripodura*) and *P.* (*Polypedilum*) is distinct, at least in the adult stage, and that subgeneric status for the two groups is warranted. Three species, *P.* (*Polypedilum*) *nubeculosum*, *P.* (*P.*) *laetum,* and *P.* (*U.*) *cultellatum* posses some "*tripodura*" like characteristics, but pending a more thorough examination of the species in *P.* (*Tripodura*), especially the immatures, Townes' three subgenera are retained.

Sasa and Kikuchi (1995) designated a new subgenus, *P.* (*Uresipedilum*), to include the species in the *convictum* or *cultellatum* group, but failed to designate a type species for their new subgenus. In a revision of *P.* (*Uresipedilum*) Oyewo and Sæther (1998) designate *P.* (*U.*) *convictum* (Walker) as the type species. They selected *P. convictum* because it is more typical of the *convictum* group and the new subgenus than is *P. cultellatum*. By designating the type species, *P.* (*Uresipedilum*) is now properly attributed to Oyewo and Sæther (1998). *Polypedilum* (*Uresipedilum*) is well defined in the adult and larval stages (Oyewo and Sæther, 1998; Sæther and Sundal, 1999). Sasa and Kikuchi (1995) also elevated the subgenus *P.* (*Pentapedilum*) to generic rank, a change we have not recognized (see Preface).

Sæther and Sundal (1999) designate a new *Polypedilum* subgenus, *P. (Cerobregma),* to include species in the *ontario* group, and they include *Asheum* (Sublette and Sublette) (=*Pedionomus*) *beckae* Sublette (1964) as a *Polypedilum* subgenus. This brings the number of subgenera recognized by Sæther and his coworkers to six, including *P.* (*Polypedilum*), *P.* (*Uresipedilum*), *P.* (*Tripodura*), *P.* (*Pentapedilum*), *P.* (*Asheum*), and *P.* (*Cerobregma*) (see Authors' Note, page v).

Oliver *et al.* (1990) list *Polypedilum californicum* (Sublette, 1960: 204) and *P. parascalaenum* (Beck, 1962: 91) in the subgenus *P.* (*Polypedilum*) in their catalog of Nearctic Chironomidae. These species were described at a time when

Sublette and apparently Beck did not consider *P.* (*Tripodura*) to be a valid subgenus. However, based on the descriptions and drawings of the male hypopygia of these species, both belong in the subgenus *P.* (*Tripodura*). Subsequently, Beck (1966: 131) considered at least the Florida specimens of *P. parascalaenum* to be the same as *P.* (*T.*) *scalaenum* (Schrank). These two species are not treated in this work.

Species Groups in Subgenera *P.* (*Polypedilum*) Kieffer & *P.* (*Uresipedilum*) Oyewo and Sæther

Most of the 24 species in the subgenus *P.* (*Polypedilum*) fall into three distinct groups based primarily on adult characters. The seven species in the subgenus *P.* (*Uresipedilum*) form one species group, the *cultellatum* or *convictum* group.

Species groups in *P.* (*Polypedilum*)

***laetum* group**

P. castornama sp. nov.
P. crassum sp. nov.
P. laetum (Meigen)
P. tuberculum sp. nov.

***illinoense* group**

P. angulum sp. nov.
P. braseniae (Leathers)
P. bergi sp. nov.
P. falciforme sp. nov.
P. illinoense (Malloch)
P. nymphaeorum sp. nov.
P. ophioides Townes

***fallax* group**

P. artifer (Curran)
P. calopterum (Mitchell)
P. fallax (Johannsen)
P. ontario (Walley)
P. pedestre (Meigen)
P. prolixipartum sp. nov.
P. vibex Townes
P. walleyi Townes

Species groups in *P.* (*Uresipedilum*)

(the ***cultellatum*** or ***convictum*** **group**)

P. aviceps Townes
P. cinctum Townes
P. cultellatum Goetghebuer
P. flavum (Johannsen)
P. obtusum Townes
P. pedatum Townes
P. sulaceps Townes

Polypedilum (*P.*) *nubeculosum* (Meigen) and *P.* (*P.*) *trigonus* Townes seem to be fairly close in all three stages and may form a fourth group in *P.* (*Polypedilum*). *Polypedilum* (*P.*) *angustum* Townes and *P.* (*P.*) *atrichon* sp. nov. are closely related and somewhat isolated from the other groups. Also, *P.* (*P.*) *albicorne* (Meigen) seems to be isolated.

Polypedilum (*U.*) *cultellatum* Goetghebuer belongs in the subgenus *P.* (*Uresipedilum*) (*convictum* group) based on the adult hypopygium, but the pupa is different from *convictum* group pupae. The hypopygia of *P.* (*U.*) *sulaceps* Townes and *P.* (*U.*) *pedatum* Townes place them in the subgenus *P.* (*Uresipedilum*), but they differ from the latter in a number of other characters. Niitsuma (1992) recognized the *convictum* group as distinct within the genus based on the "boot-shaped" superior volsellae. He placed seven Japanese species in the *convictum* group, including *P.* (*U.*) *pedatum*, *P.* (*U.*) *convictum*, *P.* (*U.*) *cultellatum*, and *P.* (*U.*) *aviceps*. The immatures of *P.* (*U.*) *sulaceps* are not known. Grodhaus and Rotramel (1980) described the immature stages of *P.* (*U.*) *pedatum*, but it was not clear from these descriptions if the larvae and pupae fit well into a particular species group; although Boesel (1985) suggested that Grodhaus and Rotramel's larval description placed *P.* (*U.*) *pedatum* in the *illinoense* group.

The heterogeneity of *Polypedilum* immature stages has been widely recognized (*e.g.* Pinder and Reiss, 1983 and 1986). If species groups in the subgenus *P.* (*Polypedilum*) were based either solely on larval characters or solely on pupal characters, they would each be different from the groupings shown above. This has led to uncertainties in the taxonomy of *Polypedilum* (*e.g.* Boesel, 1985), just as it has for chironomids in general (Pinder, 1983).

The disparity between species alignments in *P.* (*Polypedilum*) based on adults, larvae, and pupae ranges from radical to minor. Both the larvae and pupae of two species in particular, *P. braseniae* and *P. ontario*, are radically different from the immature stages of other members of the species groups to which they belong, based on adult characters. The larvae and pupae of these species have several characters that could easily place them in a genus other than *Polypedilum*.

Immatures of other species in different species groups based on adults share seemingly important characters, such as the larval mentum with even height teeth and the prealar tubercle on the pupal cephalothorax. This situation sometimes led to the placement of *P. laetum* and *P. fallax* in the same group. These associations are not supported by the characters of the other life stages (Hennig, 1950). Also, other larval and pupal characters on the same stage having the disparate characters often support their placement in the dominantly adult-based species groups.

Finally, the immatures of some species have characters that would place them in different subgroups within the dominantly adult-based species groups. For example, in the *illinoense* group, subgroups based on larval, pupal, and adult characters show minor discrepancies but broad agreement (excluding the larvae and pupae of *P. braseniae*). The apparent discrepancies may be exacerbated by the lack of discriminating characters between closely related species in larvae relative to the pupae and adults. Larval characters may reflect the fact that *Polypedilum* larvae live in such diverse habitats. Interestingly, pupae which do not feed and are relatively immobile, often show better distinguishing characters between closely related species than do the larvae.

Diagnosis and Description of the Genus *Polypedilum* Kieffer

Polypedilum Kieffer 1912

Polypedilum Kieffer, 1912: 41, type species: *Polypedilum pelostolum* Kieffer, 1912 (synonym of *Polypedilum nubifer* (Skuse)); Goetghebuer, 1937: 56; Lenz, 1941: 12; Townes, 1945: 36; Freeman 1958: 266; Freeman 1961: 707; Lenz, 1962: 237; Ashe, 1981: 58; Rossaro, 1985: 4; Boesel, 1985: 245; Bidawid and Fittkau, 1995.
Pentapedilum Kieffer, 1913: 25; Townes,1945: 61; Freeman, 1958: 266.
Polypedilum (*Tripodura*) Townes, 1945: 36.
Polypedilum (*Polypedilum*) Kieffer, 1912; Townes, 1945: 49.
Polypedilum (*Uresipedilum*) Sasa and Kikuchi, 1995: 40.

Adult Male

Diagnosis—Pulvilli split into two branches, appearing deeply bifid, each lateral branch with several subbranches. Branches and subbranches bear many fine filaments (Figure 8). Eighth abdominal segment narrowed anteriorly, triangular in shape. Medial margin of gonostylus with distinct row of evenly spaced strong setae (gonostylar medial setae, Figure 9).

Description—Flagellum of antenna with 13 flagellomeres, 13th flagellomere one to over two times as long as flagellomeres one through 12 together (Figure 2). Head without frontal tubercles, ocelli (or what appear to be ocelli) located where the frontal tubercles would be. Ocelli usually present but small. Antepronotum weakly to moderately developed anteriorly, usually visible from above only in *P. nubeculosum*. Acrostichal and dorsocentral setal rows present (Figure 7). Foretarsus without setal beard. Apical scale of foretibia either rounded apically and usually with a small spine, or roughly triangular in shape (Figures 21-24). Mesotibia with apical spine on inner comb and hindtibia with apical spine on outer comb. Wing membrane of species in subgenus *P.* (*Pentapedilum*) with setae; wing membrane of species in subgenera *P.* (*Polypedilum*), *P.* (*Tripodura*), and *P.* (*Uresipedilum*) without setae. Some species with spots on wing membrane. Squamae with setae. Hypopygium (Figure 9) with well developed anal point. Inferior volsellae long, bearing setae on dorsal surface, and usually with a posteriorly directed apical seta. Superior volsellae variable in shape (see generic diagnosis and description in Cranston *et al.*, 1989). The closely related genus *Collartomyia* can be distinguished from *Polypedilum* by the conelike anteriorly projecting scutum in *Collartomyia* (Amakye and Sæther, 1992; Amakye, 1995).

Larva

Diagnosis and Description—Mentum of four types: (1) concave medially, and with 10 or fewer, usually poorly defined, teeth (*P. braseniae* ex. *Nymphaea odorata*); (2) with 14 teeth, medial two teeth small, first lateral teeth very large and broad, these four teeth often of even height possibly due to wear (*P. braseniae* ex. *Brasenia schreberi*); (3) mentum convex, teeth 14 or 16 in number, generally of even height, medial six teeth subequal in size; (4) medial and second lateral teeth large, first lateral teeth smaller, 16 teeth present. Each of the above mentum types associated with all of following characters. Ventromental plates separated medially, without anterior crenulations, and with distinct striations except in *P.* (*Pentapedilum*) *sordens* and *P.* (*P.*) *ontario*. Antennae with five segments, some segments may be very short. Pecten epipharyngis divided into three parts, lateral parts with one to six teeth, median part with one to four teeth. Premandibles bifid, except premandibles in specimens of *P. braseniae* we examined were not bifid. Some species with small third premandibular "tooth." Segment one of maxillary palp short, less than twice as long as width at base (see generic diagnosis and description in Pinder and Reiss, 1983).

The larva of *Asheum* (=*Pedionomus*) *beckae* Sublette (1964) fits this diagnosis. This species is nearly identical to *Polypedilum* (*Pentapedilum*) *sordens* in the larval stage. The genera *Phaenopsectra* and *Endochironomus* are similar to *Polypedilum* in the larval stage, but they can be separated by differences in the heights of the mentum teeth. The mentum of *Collartomyia* larva is similar to that of *P. ontario* (Amakye and Sæther, 1992). Because the mentum of *P. ontario* is atypical for the genus *Polypedilum*, species of *Collartomyia* may be distinguishable from the more typical *Polypedilum* species based on the shape and relative height of the mentum teeth.

Pupa

Diagnosis—Most *Polypedilum* can be distinguished from other genera by the combination of all of the following characters: cephalic tubercles absent (except *P. ontario* and *P.* (*Tripodura*) *simulans*); unbroken uniserial row of hooks on posterior margin of tergite II; three lateral taeniae on segments V and VI; anal spur present on posterolateral corner of segment VIII (absent in *P. braseniae*); each anal lobe with more than 16 anal taeniae which are uniserial at least along the anterior two thirds of margin. This diagnosis will exclude several species of *Polypedilum*. Additional exceptions are mentioned in the following description. The thoracic horns in the closely related genus *Collartomyia* have many more branches than the thoracic horns of *Polypedilum* pupae.

Description—(Exuviae)

Coloration: Dorsum of cephalothorax, proximal one half to two thirds of antennal sheath, proximal portion of foreleg sheath, and margin of wing sheath brown to light brown; frontal point light brown in some species. Abdomen without pigmentation (except for *P. fallax*), except along lateral margins of segments VII and VIII which are brown to light brown. Brown color may extend to lateral margins of more anterior segments in darker specimens, color usually becomes darker posteriorly to anal spur on posterolateral corner of segment VIII. Anal spur brown to dark brown. Anal lobes stramineous to brown.

Cephalothorax: Setae described in morphology section. Cephalic tubercles absent except in *P.* (*Tripodura*) *simulans* group and *P. ontario.* Base of thoracic horn oval or circular shaped, branches of thoracic horn usually not visible on slide mounted exuviae.

Abdomen: Pedes spurii A on segment IV and pedes spurii B on segments I and II present on most species. Tergite II with prominent, unbroken, uniserial row of hooks on posterior margin. Tergites II through VI usually with multiserial row of large spinules on anterior margin, anterior spinules only slightly larger than shagreen spinules in the subgenus *P.* (*Uresipedilum*). Intersegmental spinules between segments III-IV and IV-V usually present, lateral margins of segments V and VI with three taeniae, taeniae small in *P. braseniae.* Lateral margins of segments VII and VIII with four taeniae except in *P. bergi* which may have less. Posterolateral corner of segment VIII with anal spur, except in *P. braseniae.* Each anal lobe usually with at least 16 anal taeniae. Anal taeniae uniserial along at least anterior two thirds of margin except in *P. fallax.* Anal taeniae in *P. bergi* and *P. braseniae* usually weakly multiserial along entire margin and *P. nubeculosum* may be multiserial along posterior one half of margin in some specimens (see Pinder and Reiss, 1986)

Diagnosis of Adult Males of the Subgenera *P.* (*Polypedilum*) Kieffer and *P.* (*Uresipedilum*) Oyewo and Sæther

Polypedilum (*Polypedilum*) Kieffer

P. (Polypedilum) Kieffer, 1912: 41; Townes, 1945: 61; Freeman, 1958: 267 (complete synonymy).

Diagnosis—Species of *P.* (*Polypedilum*) are distinguished from *P.* (*Pentapedilum*), *P.* (*Tripodura*), and *P.* (*Uresipedilum*) by the following. Superior volsella long and narrow, pointed or bluntly pointed apically, apical end usually curved medially, with 1 (very rarely 2) apical seta (*e. g.* Figure 9). Wings without setae. Anal point without notches along the sides and without shoulders or points on each side at the base (hypopygia with these features are shown in Figures 40-42 on page 25). *P.* (*Tripodura*) *acifer* has a sickle-shaped superior volsella with single apical seta but ninth tergite has points at base of anal point (see Figure 40 on page 25).

Polypedilum (Uresipedilum) Oyewo and Sæther

P. (Uresipedilum) Oyewo and Sæther, 1998: 317 (designation of type species); Sasa and Kikuchi, 1995: 114; Sasa and Okazawa, 1991: 54 (reference not seen, *nomen nudum, fide* Oyewo and Sæther, 1998).

Diagnosis—Species of *P.* (*Uresipedilum*) are distinguished from *P.* (*Polypedilum*), *P.* (*Pentapedilum*), and *P.* (*Tripodura*) by the following. Superior volsella with a pointed posteromedial process (boot-shaped; *e. g.*, Figures 67, 81, 82, and 123). Superior volsella with single apical seta. Wings without setae. Anal point without notches along the sides and without shoulders or points on each side at the base.

Key to Subgenera of *Polypedilum* Kieffer—Adult Males

1. Wing membrane with numerous setae (macrotrichia) .... ***P.* (*Pentapedilum*)**
1'. Wing membrane without setae **2**

2. (1'). Superior volsella (SVo) with 1 seta beyond base (rarely with 0 or 2), (as in Figures 75, 81, 86, 92) **3**
2'. SVo with several apical setae (Figures 41, 42, 67) **5**

3. (2). Apex of tergite IX with small point on either side of anal point base; anal point broad; SVo narrow, pointed apically (Figure 40) .... ***P.* (*Tripodura*) *acifer***
3'. Apex of tergite IX without small point on either side of anal point base; anal point usually narrow as in Figure 99; if anal point broadened, SVo rounded apically and bearing medial digitiform process (Figures 48, 65, 109) **4**

4. (3'). SVo rounded apically, with a posteromedial digitiform process, boot shaped, as in Figures 48, 82, 123 .... ***P.* (*Uresipedilum*)**

4'. SVo narrow, pointed apically, horn or sickle shaped, as in Figures 75, 86, 99 .... ***P.* (*Polypedilum*)**

5. (2'). SVo with medial digitiform process (Figure 67) .... ***P.* (*Uresipedilum*) *cultellatum*** (page 63)

5'. SVo without medial digitiform process, either pad-like (Figure 41), or entire SVo digitiform (Figure 42) .... ***P.* (*Tripodura*)**

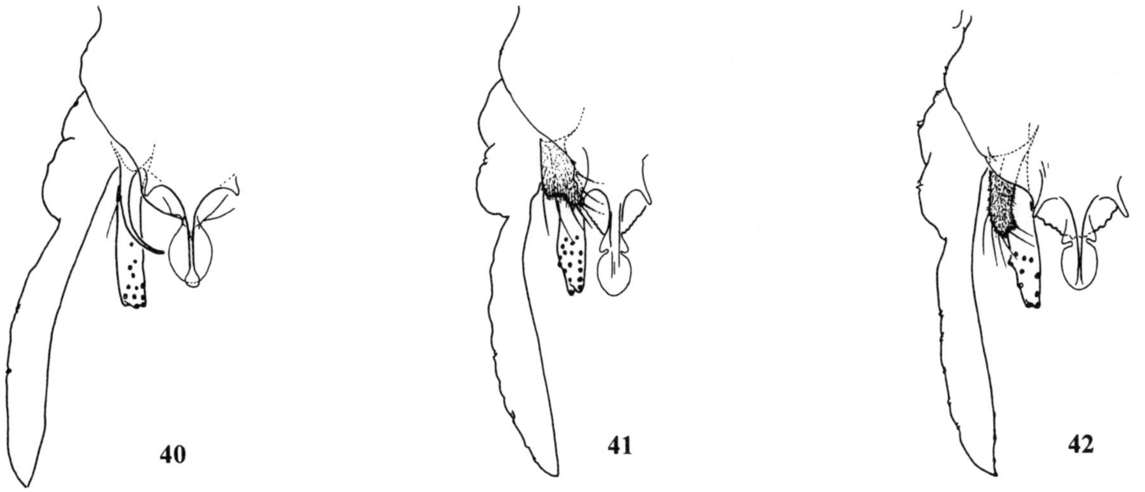

40 41 42

Figures 40-42. 40-*P.* (*Tripodura*) *acifer*, hypopygium; 41-*P.* (*Tripodura*) *simulans*, hypopygium; 42-*P.* (*Tripodura*) *digitifer*, digitiform superior volsella.

Key to Species of *P.* (*Polypedilum*) Kieffer and *P.* (*Uresipedilum*) Oyewo and Sæther—Adult Males

1. Superior volsella (SVo) rounded apically, with a posteromedial digitiform process (Figures 48, 82, 123) **2**

1'. SVo narrow, horn or sickle shaped, pointed apically **8**

2. (1). SVo with 1 seta beyond base **3**

2'. SVo with several apical setae (Figure 67) .... ***P.* (*U.*) *cultellatum*** (page 63)

3 . (2). Thorax (except scutellum), and abdominal tergites uniformly light brown to brown **4**

3'. Thorax and abdominal tergites uniformly light in color or only partially light brown **5**

4. (3). Gonostylar medial setae strong (Figure 123); scutellum lighter in color than rest of thorax; AR less than 1.25 .... ***P.* (*U.*) *pedatum*** (page 106)

4'. Gonostylar medial setae weak (Figure 127); scutellum not lighter than rest of thorax; AR greater than 1.25 .... ***P.* (*U.*) *sulaceps*** (page 113)

5. (3'). Central portion of abdominal tergites I to VI light brown; inferior volsellae (IVo) elongated posteriorly (Figure 65) .... ***P.* (*U.*) *cinctum*** (page 59)

5'. Abdominal tergites I to VI uniformly stramineous; IVo not elongated posteriorly (Figures 48, 81, 110, 111) **6**

6. (5'). Anal point very broad; SVo without apical tubercle; IVo relatively small, rounded apically (Figure 48) .... ***P. (U.) aviceps*** (page 45)

6'. SVo with or without apical tubercle; anal point narrow to moderately broadened, if very broad then SVo with distinct apical tubercle bearing distal seta; IVo larger, usually tapered apically **7**

7. (6'). SVo with or without apical tubercle, if present only apical one-half of anal point slightly broadened (Figures 81, 82) .... ***P. (U.) flavum*** (page 74)

7'. SVo with apical tubercle; anal point slightly to very broad all along its length (Figures 109-111) .... ***P. (U.) obtusum*** (page 96)

8. (1'). Abdominal tergites II-V each with a broad transverse brown band bearing numerous long setae; anal point dark, broad proximally, narrowed apically; gonostylus small (Figure 114) .... ***P. (P.) ontario*** (page 99)

8'. Without the above combination of characters **9**

9. (8'). Thorax uniformly brown to dark brown **10**

9'. Thorax uniformly stramineous or part stramineous and part light brown **25**

10. (9). Abdominal tergites uniformly brown to dark brown **11**

10'. At least abdominal tergites II and III, and usually tergites II-V part stramineous and part brown in color; or abdominal tergites II-V entirely stramineous **18**

11. (10). Posterior part of SVo base with distinct setate tubercle bearing at least 2 setae (Figures 63, 64, 133, 134) **12**

11'. SVo base without setate tubercle posteriorly (Figures 43, 56, 66, 92-94, 99, 128, 129), if small tubercle present, bearing a single apical seta (Figure 92) **13**

12. (11). Base of SVo with single seta (dashed line in Figure 63) in addition to setae on tubercle (Figures 63, 64) .... ***P. (P.) castornama* sp. nov**. (page 57)

12'. Base of SVo without setae, setae on tubercle often set on small secondary tubercles (Figures 133, 134) .... ***P. (P.) tuberculum* sp. nov.** (page 119)

13. (11'). SVo basal setae borne on small medial tubercle or on individual tubercles; gonostyli and SVo large and stout (Figure 66); foretibial apical scale without spine (Figure 21) .... ***P. (P.) crassum* sp. nov.** (page 61)

13'. Without the above combination of characters **14**

14. (13'). Antepronotum with lateral setae (Figure 7); antepronotum distinct, projecting beyond anterior margin of scutum dorsally; large species; hypopygium as in Figure 99 .... ***P. (P.) nubeculosum*** (page 87)

14'. Antepronotum without lateral setae and with antepronotum not projecting beyond anterior margin of scutum dorsally **15**

15. (14'). Wing with weak to distinct spots; hypopygium as in Figures 92-94 .... ***P. (P.) laetum*** (page 83)

15'. Wing without spots; hypopygium as in Figures 43, 56, 128 .... **16**

16. (15'). Apices of inferior volsellae (IVo) extending well beyond apex of anal point posteriorly (Figure 56); brachiolum of wing usually with 2 setae; very dark specimens of: .... ***P. (P.) braseniae*** (page 53)

16'. Apex of anal point about even with, or extending beyond, apices of IVo posteriorly (Figures 43, 128, 129); brachiolum with 1 seta **17**

17. (16'). Scutal fossal setae present (Figure 19); distal seta on SVo placed apically, SVoR less than 1.0 (Figure 43); AR less than 1.50 .... ***P. (P.) albicorne*** (page 35)

17'. Scutal fossal setae absent (Figure 20); distal seta on SVo placed near base, SVoR greater than 1.0 (Figures 128, 129); AR greater than 1.50 .... ***P. (P.) trigonus*** (page 115)

18. (10'). At least abdominal tergites II and III, and usually also tergites II-IV part stramineous and part brown in color **19**

18'. Abdominal tergites II-V entirely stramineous **21**

19. (18). Abdominal tergites II and III (usually II-V) brown except for light patch at posterolateral corner; IVo not unusually long (Figures 128, 129) .... ***P.* (*P.*) *trigonus*** (page 115)

19'. Anterior part of abdominal tergites II-V brown, posterior part stramineous; IVo very long (Figures 125, 138) **20**

20. (19'). Foretibial apical scale narrowed apically, without separate apical spine (Figure 23); SVo long, narrow as in Figure 138 .... ***P.* (*P.*) *vibex*** (page 122)

20'. Foretibial apical scale rounded apically, with or without apical spine (Figure 24); SVo shorter, stouter, as in Figure 125 .... ***P.* (*P.*) *prolixipartum* sp. nov.** (page 110)

21. (18'). Abdominal tergites VI-VIII and hypopygium light brown to brown **22**

21'. Entire abdomen stramineous; hypopygium, Figure 139 .... ***P.* (*P.*) *walleyi*** (page 124)

22. (21). Foretibia with extensive brown coloration **23**

22'. Foretibia entirely stramineous or only the apical 0.15 slightly darker **24**

23. (22). Proximal three-fourths of forefemur brown, color darkest in midpart of forefemur; haltere light; hypopygium, Figure 75 .... ***P.* (*P.*) *fallax*** (page 69)

23'. Apical one-half of foretibia and haltere brown to light brown; western species; hypopygium, Figure 124 .... ***P.* (*P.*) *pedestre*** (page 108)

24. (22'). SVo long, as in Figure 75 .... ***P.* (*P.*) *artifer*** (page 42)

24'. SVo shorter, stout, as in Figure 125; very light specimens of: .... ***P.* (*P.*) *prolixipartum* sp. nov.** (page 110)

25. (9'). SVo very large, basal setae in two groups (Figure 120) .... ***P.* (*P.*) *ophioides*** (page 102)

25'. SVo small to large but basal setae always in one group **26**

26. (25'). IVo narrow, about 5 times as long as wide (Figures 46, 47) **27**

26'. IVo wider, usually not more than 3.5 times as long as wide (Figures 44, 52, 56, 69, 70, 86, 87, 105) **28**

27. (26). Base of SVo with distinct microtrichia; apex of SVo usually with small rounded knob (Figure 46; does not show apical knob on SVo); foretibial apical scale with an apical spine .... ***P.* (*P.*) *angustum*** (page 40)

27'. Base of SVo without microtrichia, apex of SVo tapered to blunt point (Figure 47); foretibial apical scale without apical spine .... ***P.* (*P.*) *atrichon* sp. nov.** (page 43)

28. (26'). Posterior margin of abdominal tergites II-V (usually tergite I also) tinged with brown .... ***P.* (*P.*) *illinoense*** (page 78)

28'. Abdominal tergites all stramineous **29**

29. (28'). SVo usually bent sharply medially at origin of distal seta; gonostylar medial setae weak, distinctly weaker than Svo distal setae; IVo without posteriorly directed seta (Figure 44) .... ***P.* (*P.*) *angulum* sp. nov.** (page 37)

29'. Without above combination of characters **30**

30. (29'). SVo usually only slightly curved medially except near apex; IVo without posteriorly directed seta (Figures 86, 87); scutal fossal setae present; light specimens of: .... ***P.* (*P.*) *illinoense*** (page 78)

30'. Without above combination of characters **31**

31. (30'). Antennal ratio greater than 1.60 .... **32**

31'. Antennal ratio less than 1.60 .... **33**

32. (31). Phallapodeme straight, only slightly expanded at medial end (Figure 56); brachiolum of wing usually with 2 setae .... ***P.* (*P.*) *braseniae*** (page 53)

32. Phallapodeme expanded at medial end (Figure 52); brachiolum with a single seta .... ***P.* (*P.*) *bergi* sp. nov.** (page 49)

33. (31'). SVo usually distinctly narrowed apically to point of distal seta origin, usually only slightly curved medially; phallapodeme long, not expanded at medial end; gonostylar medial setae distinctly weaker than SVo distal seta (Figure 105) .... ***P.* (*P.*) *nymphaeorum* sp. nov.** (page 92)

33'. SVo evenly tapered to sharp point, curved medially toward anal point; phallapodeme expanded at medial end; gonostylar medial setae stronger (Figures 69, 70) .... ***P.* (*P.*) *falciforme* sp. nov.** (page 66)

Key to Species of *Polypedilum* Kieffer—Larvae

(The larvae of some species could not be separated, in which case more than one species keys to the same couplet).

1. Mentum either as in Figure 57 or Figure 58; three parts of pecten epipharyngis (PE) without teeth (Figure 61); premandibles simple (Figure 60) or bifid .... ***P.* (*P.*) *braseniae*** (page 54)

1'. Mentum not as in Figures 57 or 58; at least lateral parts of PE with 2 or more teeth; premandibles always at least bifid **2**

2. (1'). Teeth of mentum of about equal height (Figures 25, 76, 95, 116, 135) **3**

2'. First lateral teeth of mentum much smaller than medial 2 teeth and second lateral teeth (Figure 100) **8**

3. (2). Head triangular shaped; mouth parts unusually small; striations on ventromental plate not distinct (Figures 115, 116) .... ***P.* (*P.*) *ontario*** (page 101)

3'. Head and mouth parts of normal shape and size; striations distinct, if not distinct ventromental plates as in Figure 25 .... **4**

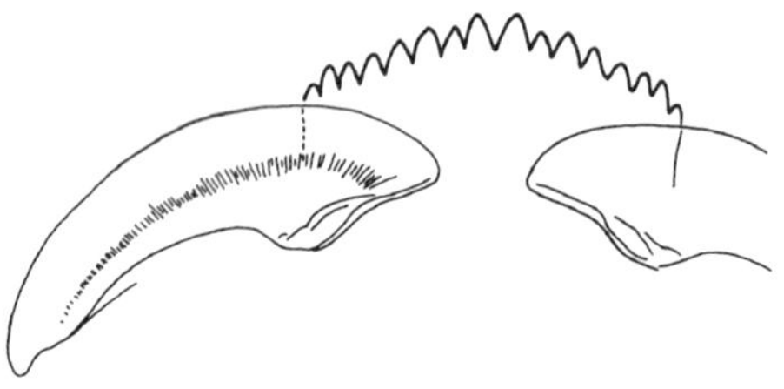

Figure 25. *P.* (*Pent.*) *sordens* (v.d. Wulp), mentum.

4. (3'). Ventromental plate very broad, width 5 times the distance between ventromental plates, striations not distinct, visible only in medial part (Figure 25) .... ***P.* (*Pent.*) *sordens***

4'. Ventromental plate not as broad, width no more than 4 times the distance between ventromental plates, striations more distinct **5**

5. (4'). Ventromental plate with well developed posterior lobe (Figure 49); antennal segments 1 and 2 very long, segment 2 about twice as long as segments 3, 4, and 5 together (Figure 50) .... ***P.* (*U.*) *aviceps*** (page 47)

5'. Ventromental plates with poorly to moderately well developed posterior lobe (Figures 95, 135, 76); antennal segments 1 and 2 not unusually long, segment 2 about same length, or considerably shorter than segments 3, 4, and 5 together (Figures 77, 96, 136) **6**

6. (5'). Mean width of ventromental plates no more than 2.5 times distance between ventromental plates (Figures 95, 135); antennal segments 3, 4, and 5 together considerably longer than segment 2 ... 7

6'. Mean width of ventromental plates at least 2.5 times distance between plates (Figure 76); antennal segments 3, 4, and 5 about as long as segment 2 (Figure 77) .... ***P. (P.) fallax*** (page 71)

7. (6). Ring organ on antennal segment 1 located near base, ROR greater than 3.5 (Figure 96) .... ***P. (P.) laetum*** (page 84)

7'. Ring organ located more distally, ROR less than 3.5 (Figure 136); mentum, Figure 135 .... ***P. (P.) tuberculum* sp. nov.** (page 120)

8. (2'). Antennal segment 3 short, one-third or less length of segment 2, or if longer, ventromental plates with well developed posterior lobes (Figure 83) **9**

8'. Antennal segment 3, one-half or more length of segment 2, or if shorter, ventromental plates without posterior lobes **14**

9. (8). Ventromental plates without posterior rounded lobes (Figure 26) **10**

9'. Ventromental plates with well developed posterior lobes (Figures 49, 83) **12**

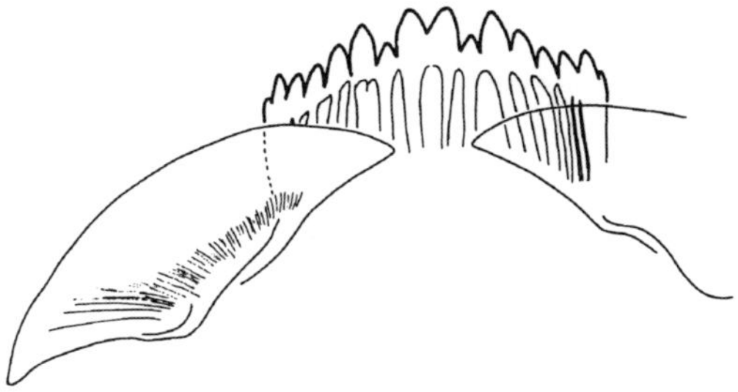

Figures 26. *P.* (*T.*) *digitifer* Townes, mentum.

10. (9). Antennal segments 3, 4, and 5 together much shorter than segment 2 (Figure 27) .... ***P. (T.) scalaenum*** or ***P. (T.) albinodus*** **?**

10'. Antennal segments 3, 4, and 5 together about as long as segment 2; antennal segment 4 much longer than segment 3 (Figure 28) **11**

Figures 27-28. 27-*P.* (*T.*) *scalaenum* (Schrank), antenna, segments 2-5; 28-*P.* (*T.*) *simulans* Townes, antenna, segments 2-5.

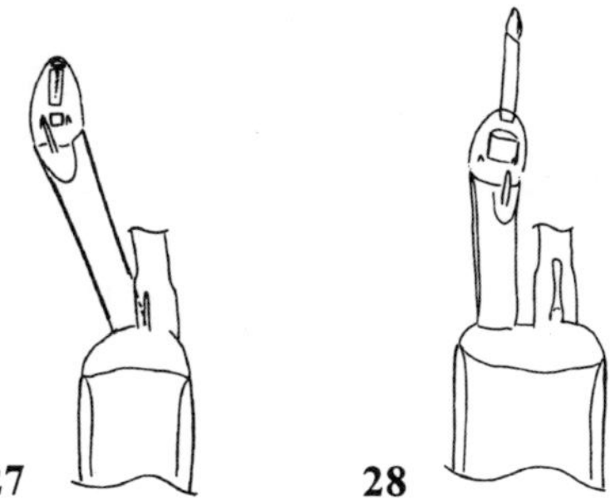

27 28

11. (10'). Lateral parts of PE with 3 teeth (based on single specimen) .... ***P. (T.) halterale***

11'. Lateral parts of PE with at least 4 teeth .... ***P. (T.) simulans*** or ***P. (T.) digitifer***

12. (9'). Antennal segment 2 about twice as long as segments 3, 4, and 5 together; first lateral teeth of mentum often only slightly shorter or equal to height of second lateral teeth (Figure 49) .... ***P. (U.) aviceps*** (page 47)

12'. Antennal segment 2 at most 1.5 times as long as segments 3, 4, and 5 together; first lateral teeth much shorter than second lateral teeth (Figure 83) (These two species are possibly separable using couplet 13) .... ***P. (U.) flavum*** (page 76) or ***P. (U.) obtusum*** (page 97)

13. (12'). Lauterborn organs exceeding apex of antennal segment 3 (Figure 112) .... **P. (U.) obtusum** (page 97)

13'. Lauterborn organs not exceeding apex of antennal segment 3 (Figure 84) .... **P. (U.) flavum** (page 76)

14. (8'). Width of ventromental plate greater than 2.5 times distance between ventromental plates **15**

14'. Width of ventromental plate 2.5 times or less distance between ventromental plates **17**

15. (14). Labral seta I broad apically, with apical border of fine, short filaments (Figure 102); antenna as in Figure 101 .... **P. (P.) nubeculosum** (page 89)

15'. Labral seta I lanceolate, with lateral fringe of long, fine filaments as in Figure 97; antennae as in Figures 29, 131 .... **16**

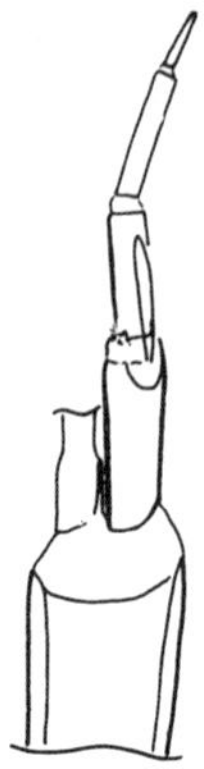

29

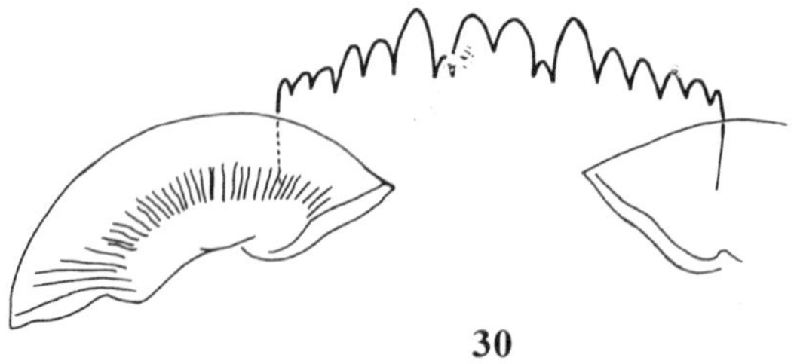

30

Figures 29-30. *P.* (*Pent.*) *tritum* (Walker): 29-antenna; 30-mentum.

16. (15'). Primary antennal blade exceeding apex of segment 5 apically; Lauterborn organs distinct, extending nearly to apex of segment 3 (Figure 131); ventromental plates 2.75 times or greater distance between ventromental plates (Figure 130) .... **P. (P.) trigonus** (page 117)

16'. Primary antennal blade reaching only the base of segment 5; Lauterborn organs not distinct, smaller (Figure 29); ventromental plates usually less than 2.5 times distance between ventromental plates, but variable (Figure 30) .... **P. (Pent.) tritum**

17. (14'). Median two teeth of mentum distinctly shorter than second lateral teeth (Figure 53) .... **P. (P.) bergi sp. nov.** (page 51)

17'. Median two teeth of mentum equal in length or possibly a little longer than second lateral teeth **18**

18. (17'). Ventromental plate as in Figure 30; Lauterborn organs indistinct, small or absent .... **P. (Pent.) tritum**

18'. Ventromental plates as in Figures 71, 88, 106; Lauterborn organs distinct, reaching mid part of antennal segment 3 or beyond **19**

(The following species are difficult to separate in the larval stage).

19. (18'). Antennal segments 3 and 4 about equal in length, or 4 longer than 3; if 3 slightly longer than 4, width of ventromental plate 1.7 times or less distance between ventromental plates **20**

19'. Antennal segment 3 longer than 4; if about equal in length than width of ventromental plates greater than 1.7 times distance between ventromental plates **22**

20. (19). Lauterborn organs reaching to apex of antennal segment 3 (Figure 121).. ***P. (P.) ophioides*** (page 104)

20'. Lauterborn organs reaching just beyond the mid part of antennal segment 3, or Lauterborn organs indistinct **21**

21. (20'). Antennal segments 3 and 4 about equal in length (Figure 89) .... ***P. (P.) illinoense*** (page 81)

21'. Antennal segment 3 usually distinctly shorter than segment 4 (Figure 45) .... ***P. (P.) angulum*** **sp. nov.** (page 39)

22. (19'). Antennal segment 3 consistently longer than segment 4; width of ventromental plate 1.44-1.77 times distance between ventromental plates (Figure 71) .... ***P. (P.) falciforme*** **sp. nov.** (page 68)

22'. Antennal segment 3 may or may not be longer than segment 4; width of ventromental plate 1.75-1.95 times distance between ventromental plates (Figure 106)..... ***P. (P.) nymphaeorum*** **sp. nov.** (page 94)

Key to Species of *Polypedilum* Kieffer—Pupae

(The pupae of some species could not be separated. In this case, more than one species keys to the same couplet).

1. Cephalic tubercles present (Figures 31, 32, 118) **2**

1'. Cephalic tubercles very small or absent (Figure 17) **3**

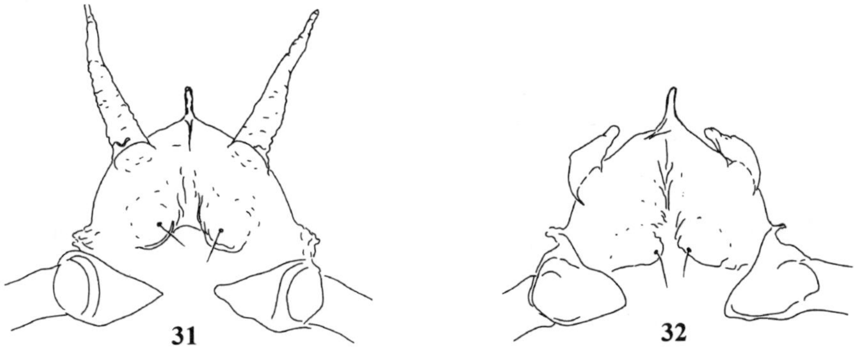

Figures 31-32. *P.* (*T.*) *simulans* Townes: 31-cephalic tubercles; 32-cephalic tubercles.

2. (1). Cephalic tubercles broadly based, cone shaped and darkened at tip (Figure 118); abdominal tergites without anterior row of large spinules (Figure 117) .... ***P. (P.) ontario*** (page 102)

2'. Cephalic tubercles long, narrow, not darkened at tip (Figures 31, 32); abdominal tergites with anterior row of large spinules (Figure 37) .... ***P. (T.) simulans***

3. (1'). Abdominal tergite VIII without posterolateral anal spur; taeniae on lateral margins of segments V and VI small (Figure 62) .... ***P. (P.) braseniae*** (page 56)

3'. Abdominal tergite VIII with posterolateral anal spur; taeniae on segments V and VI of normal size.. **4**

4. (3'). Cephalothorax with distinct tubercle anterior to base of wing sheath (Figure 74) **5**

4'. Cephalothorax without such a tubercle **7**

5. (4). Tergite II nearly completely covered with spinules and all spinules relatively large and prominent (Figure 68) .... ***P. (U.) cultellatum*** (page 65)

5'. Spinules on tergite II less extensive; spinules generally smaller and weaker (Figures 55, 73) **6**

6. (5'). Fewer than 40 anal taeniae on each anal lobe; smaller species, total length of pupal exuviae less than 5 mm (Figure 73) .... ***P. (P.) falciforme* sp. nov.** (page 68)

6'. Forty or more anal taeniae on each anal lobe; larger species, total length of pupal exuviae 5 mm or more (Figure 55) .... ***P. (P.) bergi* sp. nov.** (page 52)

7. (4'). Intersegmental spinules between segments III and IV absent (Figures 33, 51, 85, 113, 132) **8**

7'. Intersegmental spinules between segments III and IV present (Figures 79, 90, 103) **12**

8. (7). Anterior row of spinules on tergite II multiserial, large and prominent (Figures 132, 33) **9**

8'. Anterior row of spinules on tergite II nearly absent to multiserial but very weak, spinules at most only slightly larger than shagreen spinules (Figures 51, 85, 113) **10**

9. (8). Shagreen pattern on tergites IV and V in 4 separate patches (Figure 33) .... ***P. (Pent.) sordens***

9'. Shagreen pattern on tergites IV and V with medial longitudinal row usually connecting anterior row with posterior spinules (including intersegmental spinules between IV and V) (Figure 132) . ... ***P. (P.) trigonus*** (page 117)

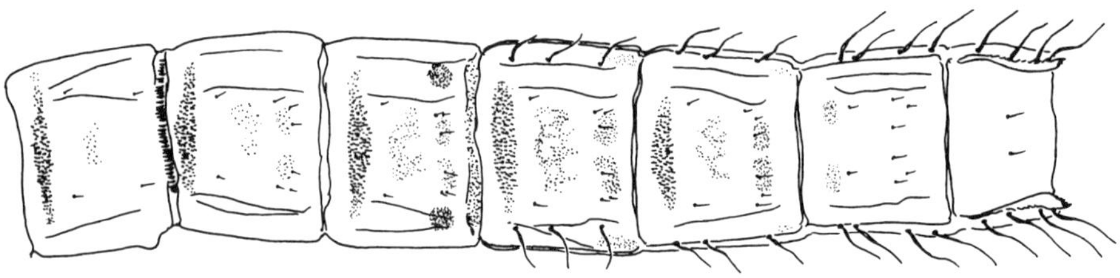

Figure 33. *P.* (*Pent.*) *sordens* (v.d. Wulp) pupal abdomen.

10. (8'). Each anal lobe with fewer than 30 anal taeniae .... ***P. (U.) aviceps*** (page 48)

10'. Each anal lobe with 30 or more anal taeniae **11**

11. (10'). Pedes spurii B usually absent on segment II; midline of tergites III-VI usually free of spinules, except posteriorly (Figure 85) ... . ***P. (U.) flavum*** (page 76)

11'. Pedes spurii B present on segment II; spinule pattern on tergites III-VI not separated by bare midline (Figure 113) .... ***P. (U.) obtusum*** (page 98)

12. (7'). Base of anal spur, or posterolateral margin of tergite VIII with many small lateral accessory points in addition to apical accessory points (Figures 80, 104, 119) **13**

12'. Base of anal spur, or posterolateral margin of tergite VIII without small lateral accessory points; apical accessory points of variable size usually present (Figures 34, 35) **18**

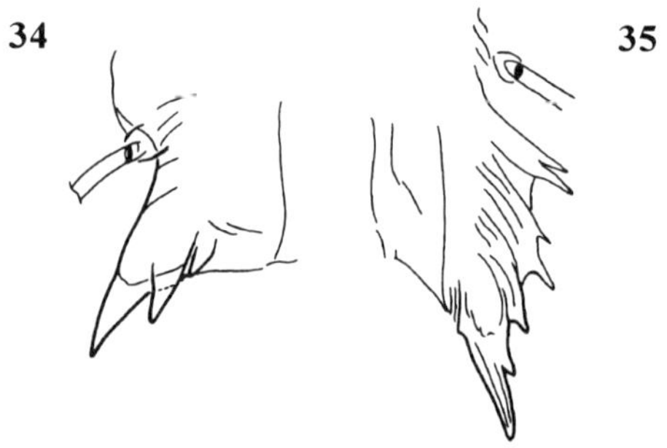

Figures 34-35. 34-*P. falciforme* sp. nov., anal spur on tergite VIII; 35-*P.* (*T.*) *digitifer* Townes, anal spur on tergite VIII.

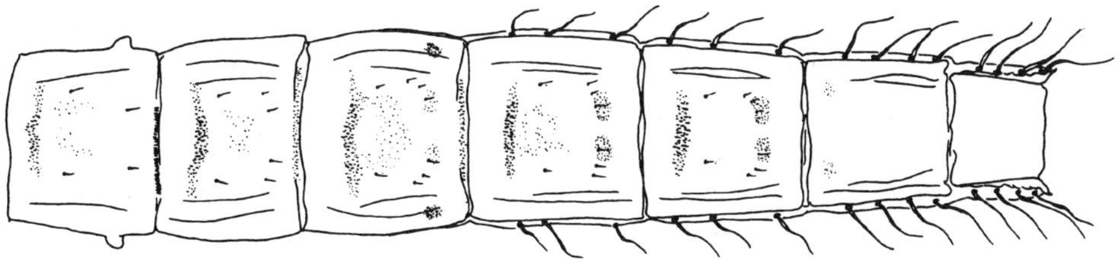

Figure 36. *P.* (*Pent.*) *tritum* (Walker), pupal abdomen.

13. (12). Shagreen spinule pattern on all tergites, especially tergites II and III, generally more extensive (Figures 79, 98, 137); if somewhat sparse, shagreen pattern triangular shaped in central parts of tergites III and IV (Figure 137) **14**

13'. Shagreen pattern generally more sparse (Figures 36, 103, 126) **16**

14. (13). Shagreen pattern with large fenestrations (Figures 98, 137) **15**

14'. Fenestrations very small or absent; posterior tergites usually tinged with brown (Figure 79) .... ***P.* (*P.*) *fallax*** (page 72)

15. (14). Shagreen spinules very large, about equal in size to hooks on tergite II; shagreen pattern covering most of tergites II and III (Figure 98) .... ***P.* (*P.*) *laetum*** (page 86)

15'. Shagreen spinules smaller than hooks on tergite II; shagreen pattern triangular shaped on tergites II-IV (Figure 137) .... ***P.* (*P.*) *tuberculum* sp. nov.** (page 121)

16. (13'). Shagreen pattern without fenestrations; small species, total length of pupal exuviae usually less than 4 mm (Figure 36) .... ***P.* (*Pent.*) *tritum***

16'. Shagreen pattern with large fenestrations; larger species, total length of pupal exuviae usually over 5 mm (Figures 103, 126) **17**

17. (16'). Shagreen spinules on tergites IV and V separated into 4 patches, posterior two distinctly separate (Figure 103) .... ***P.* (*P.*) *nubeculosum*** (page 90)

17'. Shagreen spinules on tergites IV and V separated into 3 patches, posterior two merging, or at least touching medially (Figure 126) .... ***P.* (*P.*) *prolixipartum* sp. nov.** (page 112)

18. (12'). Anterior row of large spinules on tergites III-VI distinctly larger than hooks on posterior margin of tergite II, and intersegmental spinules (Figures 90, 122) **19**

18'. Anterior row of large spinules on tergites III-VI about the same size or smaller than hooks on posterior margin of tergite II (Figures 37-39, 108), if somewhat larger, shagreen pattern triangular shaped on tergites III-VI (Figure 108) **20**

19. (18) Pedes spurii A on tergite IV absent; fenestrations few in number or absent (Figure 122) .... ***P.* (*P.*) *ophioides*** (page 105)

19'. Pedes spurii A on tergite IV present; fenestrations usually present on tergites III-V (Figure 90) .. .. ***P.* (*P.*) *illinoense*** (page 82) or ***P.* (*P.*) *angulum* sp. nov.** (page 39)

20. (18') Shagreen pattern in central parts of tergites III-VI triangular in shape (Figure 108) .... ***P.* (*P.*) *nymphaeorum* sp. nov.** (page 94)

20'. Shagreen pattern not triangular shaped on tergites III-VI **21**

21. (20') Shagreen pattern covering most of tergites II and III; shagreen spinules large, nearly as large as spinules in anterior rows (Figure 39) .... ***P.* (*T.*) *scalaenum*** group

21'. Shagreen spinules in two lateral patches on tergites III and IV, separated by bare midline (Figure 38) .... ***P.* (*T.*) *digitifer***

21'. Medial parts of tergites III and IV usually with spinules (Figure 37) .... ***P.* (*T.*) *simulans***
(Species close to *P.* (*T.*) *simulans* may key here).

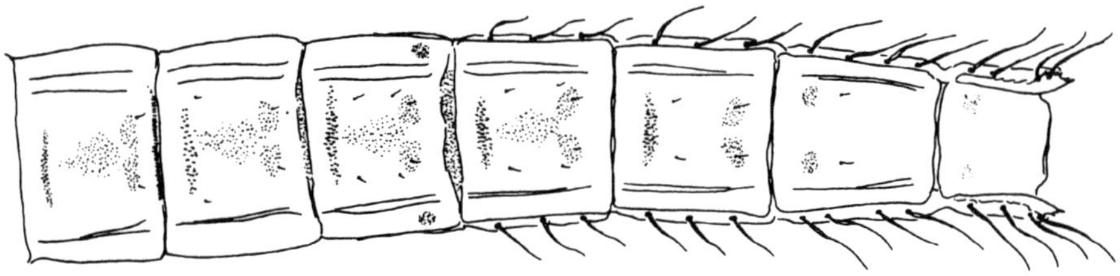

37

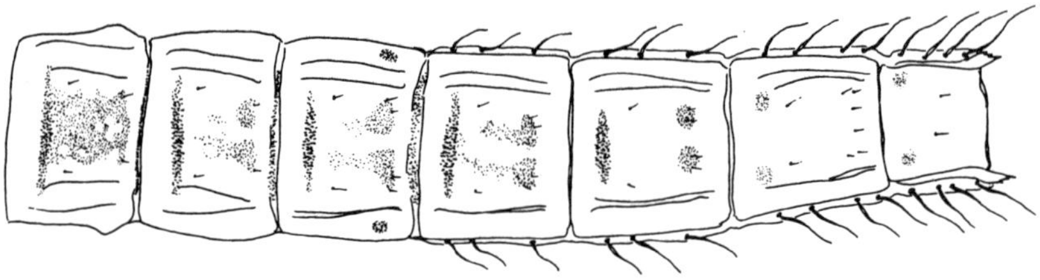

38

39

Figures 37-39. 37-*P.* (*Pent.*) *simulans* (Walker), pupal abdomen; 38-*P.* (*T.*) *digitifer* Townes, pupal abdomen; 39-*P.* (*T.*) *scalaenum* (Schrank), pupal abdomen.

Descriptions of Nearctic Species of *P.* (*Polypedilum*) Kieffer and *P.* (*Uresipedilum*) Oyewo and Sæther

Polypedilum (P.) albicorne (Meigen) 1838

Chironomus albicornis Meigen, 1838: 6; Gimmerthal, 1846: 5 ? (description indicates "hairy" wings); Edwards, 1929: 403.

Polypedilum albicorne (Meigen). Goetghebuer, 1931b: 94; Goetghebuer, 1937: 58; Brundin, 1947: 63; Brundin, 1949 (biology); Albu, 1963: 226 (figure of hypopygium); Lehmann, 1971: 500 (biology, figure of pupa); Langton, 1984: 249 (figure of pupa); Rossaro, 1985: 17 (figure of hypopygium, and larval and pupal characters); Ashe and Cranston, 1991 (Palaearctic species); Caldwell *et al.*, 1997.

Polypedilum nigritum Townes, 1945: 56; Townes, 1952: 52 (new synonymy); Beck, 1977: 192 (biology); Harper and Cloutier, 1979 (biology); Hudson *et al.*, 1990.

Adult Male

Diagnosis— Dark species, scutal fossal setae present; SVo long and thin, SVoR 1 or less; AR less than 1.50.

Description—(n=10)

Coloration: Pedicel, head, thorax and coxae brown to dark brown; abdominal tergites brown but usually lighter than thorax; palps and legs except coxae uniformly light brown to stramineous; abdominal sternites light brown; haltere stramineous to very light brown; wing veins stramineous.

Biometrics: L total 3.28 (2.65-3.60) (9) mm; L thorax 0.93 (0.77-1.01) (9) mm; L wing 1.82 (1.47-2.04) mm; W wing 0.55 (0.49-0.62) mm

Antenna: AR 1.19 (1.02-1.33) (8). D scape 141 (120-160) (9). L pedicel 106 (100-120) (5). L flagellomeres 1-12 444 (410-470) (8). L flagellomere 13 529 (420-600) (8).

Head: W 450 (410-480) (8). Interocular distance 64 (50-80) (7). HIR 7.11 (5.75-8.20) (7). Vertical S 12.4 (10-15). Clypeal S 11.9 (10-15). Ocelli very small, not visible on some specimens. L maxillary palp segments 1-5: 34 (27-41), 35 (27-41), 90 (79-104), 114 (97-131) (9), 174 (144-212) (9).

Thorax: Campaniform sensilla on dorsal edge of antepronotum small, only slightly raised from surface, not visible on some specimens. Postpronotal depression small and without visible campaniform sensilla. Antepronotum without setae. Scutal fossal S 2.6 (1-3); when 3 present, setae form triangle or lie on line roughly perpendicular to dorsocentral setal row. Dorsocentral S 10.3 (8-15), uniserial. Prealar S 4.6 (3-6). Scutellar S in 2 rows, anterior row with smaller and fewer setae. Scutellar S in anterior row 4.3 (1-8) (9), posterior row 10.7 (8-13) (9).

Wing: VR 1.27 (1.22-1.31). Wing without spots. Apex of R_{4+5} usually closer to wing apex than apex of M. R_1 and R_{2+3} adjacent along their length. Brachiolar S 1. R_1 S 32.2 (27-40). R_{4+5} S 28.5 (21-38). Squamal S 7.9 (5-12) (7).

Legs: Foretibial apical scale with spine, occasionally dark in some specimens. L scale 34 (23-43) (8). L spine 8 (5-14) (8). L mesotibial spine 22 (16-27). L hindtibial spine 26 (20-32).

	Fe	T1	Ta1	Ta2	Ta3
PI	748 (630-830) (9)	584 (480-660) (9)	924 (730-970) (9)	596 (510-700) (8)	413 (360-480) (8)
PII	860 (730-970)	730 (610-810)	395 (340-430)	251 (210-290)	185 (150-210)
PIII	944 (800-1050)	840 (700-950)	614 (540-680)	360 (320-400)	303 (260-340)

	Ta4	Ta5	LR	BV	SV
PI	289 (260-330)	120 (70-150)	1.49 (1.40-1.53)	1.52 (1.47-1.58)	1.54 (1.48-1.64)
PII	118 (100-130)	76 (60-90)	0.54 (0.53-0.56)	3.16 (3.04-3.26)	4.01 (3.76-4.14)
PIII	175 (150-190)	89 (80-100)	0.72 (0.68-0.74)	2.63 (2.55-2.71)	2.97 (2.83-3.09)

Hypopygium (Figure 43): Tergite IX S 10.7 (8-13), located in light oval-shaped area as in *P. laetum*. Anal point tapered from base to apex. Anal point usually exceeding apex of IVo, 16 (0-43). Base of SVo small, rounded without tubercles or microtrichia. Basal S 2.2 (2-3). Apical portion of SVo usually narrow and straight, angled medially or with only slight inward curve. SVo often swollen at point of distal seta origin. Distal S placed beyond middle of SVo, mean SVoR 0.62 (0.31-1.00). One specimen with no distal seta on one SVo. Max L distal S 98 (90-117). IVo generally straight, angled somewhat medially and of uniform width, apex usually rounded, apical or subapical posteriorly directed seta present. Dorsal S 11.0 (9-14). Gonostylus with rounded lateral margin, medial margin straight or sometimes slightly concave. Gonostylar medial S 3.3 (3-4) (9). Max L medial S 73 (56-81) (9). Max L apical S 29 (23-34) (8). L gonostylus 138 (119-153). GsR 3.89 (3.16-4.42). L posterior lobe gonocoxite 67 (59-77). HR 2.07 (1.87-2.24). HV 2.40 (2.23-2.61). Phallapodeme expanded at point of articulation and at its medial end.

Type locality: *P. albicorne*, locality unknown, August, male, (? Paris; museum); *P. nigritum*, New York, Ithaca, 11 Sept. 1935, H. K. Townes, male, (HKT).

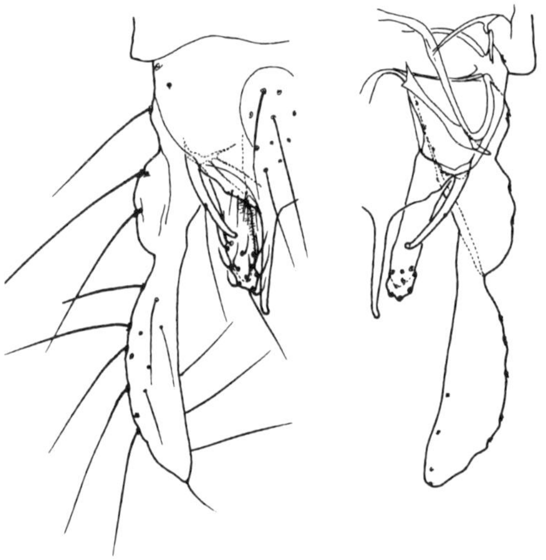

Figure 43. *P. albicorne* (Meigen), hypopygium.

Larva and Pupa
See Rossaro (1985, figures of larva and pupa) and Langton (1984, figure of pupa)

Remarks
The name *Chironomus pullus* (Zetterstedt, 1838) as applied by Goetghebuer (1921: 140), who figured the hypopygium, appears to be a synonym of *P. albicorne*. However, Edwards (1929: 405), after examining both types, suggested that they were not the same species. Lately the name has been applied to a species in the subgenus *P.* (*Tripodura*) (Brundin, 1947; Hirvenoja, 1962).

Townes (1952) synonymized his species *P. nigritum* with *P. albicorne* after looking at a specimen of the latter species from Europe. We have seen five male specimens of *P. albicorne* from the Fulda River, Germany, which conform very well to the Nearctic specimens examined. They differed only in having a somewhat more prominent dorsal part of the antepronotum, and the dorsocentral setal row extended further anteriorly nearly meeting the scutal fossal setae. We believe this to be a Holarctic species. Rossaro's (1985) figure of the hypopygium seems to support this conclusion. Rossaro (1985) has figured characters of the immatures as well, and separates them from the immatures of other Italian *Polypedilum* in his keys. We have not seen the immatures of this species.

Biology

Polypedilum albicorne is a cold stenothermic species, reared from larvae collected from moss-covered rocks in strong current (Lehmann, 1971). Beck (1977) summarized the environmental requirements of *P. nigritum*. Caldwell *et al.* (1997) in their update of Hudson *et al.* (1990) list *P. albicorne* in their checklists of species found in the southeastern United States. Harper and Cloutier (1979) reported *P. nigritum* in moderate numbers but widespread in the L'Achigan River, Quebec study reach.

Material Examined—Adult Specimens

Paratype male, **New York**, Ithaca, May, (Corn); 2 paratypes, no data (slides of hypopygia) (USNM). **Massachusetts**, fern bog, Concord, 22 July 1961, W.W. Wirth, 2 males, (USNM). **Minnesota**, 2 mi. N Hovland on U.S. Highway 61, Cook Co., June through Sept. 1968, 1970, leg. E.F. Cook, New Jersey mosquito light trap, 14 males, (UMn); U.S.EPA Lab., Duluth, St. Louis Co., 17 June 1969, leg. E.F. Cook, New Jersey mosquito light trap, (UMn). **New Hampshire**, creek margin, Jefferson Notch, 30 July 1951, W.W. Wirth, 1 male, (USNM). **New York**, Ellis Hollow, Tompkins Co., 15 June 1963, C.O. Berg, light trap, 3 males, (USNM); spring seep, Cow Neck, Long Island, 24 May 1963, W.W. Wirth, 2 males. (USNM); Allegany State Park 28 May 3 to June 1963, W.W. Wirth, light trap, 3 males, (USNM). **Virginia**, Holmes Run, Falls Church, 17 Oct. 1960, W.W. Wirth, light trap, 3 males, (USNM); Vesuvius, Rockbridge Co., July 1960, D.H. Messersmith, light trap, 1 male, (USNM). **Wisconsin**, 11 mi. E, 4 mi. S Siren, Burnett Co., 9 July 1966, D.C. Hansen, light trap, 1 male, (UMn). **British Columbia**, Hixon, June through Aug. 1966, 10 males, (CNC). **Newfoundland**, experimental farm, St. Johns, 21 July 1967, malaise trap, J.F. McAlpine, 1 male, (CNC). Records from May through October.

Polypedilum (P.) angulum New Species

Adult Male

Diagnosis—Very close to *P. illinoense* but distinguished by hypopygial differences. SVo usually bent sharply medially at point of distal seta origin. Medial margin of IVo nearly straight, subparallel to lateral margin. Phallapodeme narrow medially. Gonostylar medial setae less than 45 microns in length, distinctly weaker than SVo distal setae.

Description—(n=9)

Coloration: Generally light. Haltere, pedicel, most of thorax, wing veins, and abdomen stramineous. Dorsal part of scutellum usually tinged with brown. Dorsal vittae usually, lateral vittae less often, tinged with light brown. Few darker specimens have generally brown thorax. Flagellum light brown to brown. Legs stramineous, apical 1/6 of forefemur, foretibia and foretarsal segments 1-4 often shade darker.

Biometrics:	L total 3.26 (2.70-3.62) mm	L thorax 0.97 (0.81-1.14) mm
	L wing 1.85 (1.47-2.22) mm	W wing 0.58 (0.46-0.66) (8) mm

Antenna: AR 1.61 (1.43-1.88) (7). D scape 153 (140-170) (6). L pedicel 110 (1). L flagellomeres 1-12 436 (400-490) (7). L flagellomere 13 697 (600-760) (7).

Head: W 490 (450-520) (7). Interocular distance 80 (60-90) (7). HIR 6.13 (5.33-7.50) (7). Vertical S 11.3 (10-13) (6). Clypeal S 15.7 (14-17) (6). Ocelli apparently absent. L maxillary palp segments 1-5: 39 (34-47) (4), 42 (38-47) (4), 111 (90-133) (4), 111 (101-126) (4), 185 (153-203) (4).

Thorax: Antepronotum and thoracic sensilla as in *P. illinoense*. Scutal fossal S 1.3 (0-4). Dorsocentral S 12.5 (9-16), uniserial, 1 or 2 anterior setae often located anteriorly near scutal fossal S. Prealar S 4.6 (3-6). Scutellar S in 2 regular rows, anterior row smaller; anterior row 5.1 (2-8) (7); posterior row 8.3 (6-10) (7).

Wing: VR 1.22 (1.20-1.27), wing unspotted. Apices of R_{4+5} and M about equidistant from wing apex with R_{4+5} closer in 1 measured specimen. R_1 and R_{2+3} adjacent apically. Brachiolar S 1. R_1 S 37.3 (28-48). R_{4+5} S 30.1 (22-39). Squamal S 8.4 (6-11) (6).

Legs: L foretibial apical scale 32 (25-38). L spine on foretibial scale 9 (5-14), tip of spine dark on some specimens. Mesotibial and hindtibial apical spines similar to *P. illinoense.*

	Fe	**T1**	**Ta1**	**Ta2**	**Ta3**
PI	857 (720-990)	541 (400-650)	991 (830-1090) (7)	739 (630-830) (7)	489 (420-570) (6)
PII	933 (780-1060)	764 (640-900)	441 (380-500) (8)	284 (230-340) (8)	201 (100-240) (8)
PIII	989 (820-1140)	904 (750-1040)	664 (570-730) (7)	390 (330-450) (7)	316 (270-360) (7)

	Ta4	**Ta5**	**LR**	**BV**	**SV**
PI	342 (300-370) (6)	142 (110-160) (6)	1.78 (1.69-1.85) (7)	1.41 (1.36-1.48) (6)	1.44 (1.37-1.49) (7)
PII	128 (100-140) (8)	73 (60-90) (8)	0.58 (0.56-0.60) (8)	3.13 (2.97-3.20) (8)	3.84 (3.74-3.98) (8)
PIII	196 (160-220) (7)	91 (80-100) (7)	0.72 (0.70-0.74) (7)	2.63 (2.59-2.68) (7)	2.93 (2.80-3.02) (7)

Hypopygium (Figure 44): Tergite IX S 5.1 (2-7), located in narrow, middorsal area; usually not as triangular shaped as in *P. illinoense.* Anal point straight or curved ventrally, usually does not exceed IVo posteriorly, -11.3 (5 to -25). SVo bent sharply medially just beyond base or near point of distal seta origin; tip usually rounded and curved medially. Dorsal surface and medial margin of basal portion with sparse but strong and distinct microtrichia. Basal S 2.9 (2-4). L SVo 76 (70-86). Placement of distal S not as variable as in *P. illinoense,* mean SVoR 1.50 (1.00-1.82). Max L distal S 77 (68-90) (7), exceeding apex of IVo in some specimens. IVo angled medially, medial margin nearly straight, not curved as in *P. illinoense*; dorsal surface, except the extreme posterolateral margin, bearing microtrichia. L IVo 101 (83-115). W IVo 32 (25-47) (7). Dorsal S 10.3 (8-13), usually not as strong as in *P. illinoense.* IVo without apical posteriorly directed seta. Gonostylus as in *P. illinoense*; medial S 3.5 (3-5), distinctly weaker than SVo distal S. Max L medial S 38 (34-43) (8). Max L apical S 20 (14-25) (6). L gonostylus 157 (140-171). GsR 4.79 (4.17-5.17) (8). HR 3.22 (2.83-3.62). L posterior lobe gonocoxite 49 (47-54). HV 2.10 (1.88-2.37). Phallapodeme usually curved near its midpart, medial end not expanded.

Type locality: Holotype male, **Minnesota**, Minnesota Forestry Station, Little Fork, Koochiching Co., 26 Aug. 1970, leg. E.F. Cook, New Jersey mosquito light trap, (UMn). Paratypes are designated as indicated in material examined.

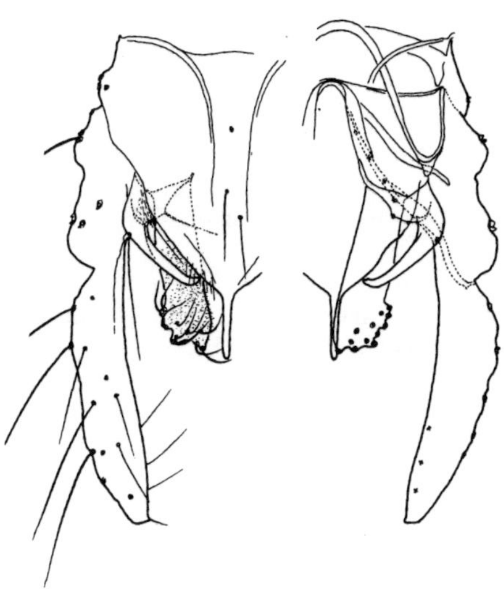

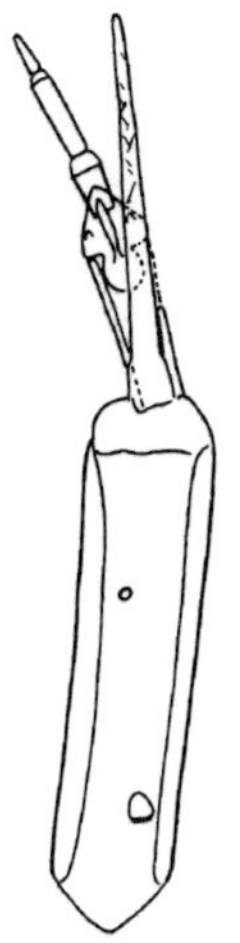

Figures 44-45. *P. angulum* sp. nov.: 44-hypopygium; 45-antenna.

Larva

Diagnosis—Very close to *P. illinoense* and probably cannot be consistently distinguished from this species in larval stage. Antennal segment 4 longer than 3 in *P. angulum*. Segments 3 and 4 same size or 3 longer than 4 in *P. illinoense*. PS on second segment shorter. Segment 3 broadened apically.

Description—(n=7, all males)

Biometrics: L head capsule 160 (153-169)

Mentum: W 89 (77-104) (6), otherwise as in *P. illinoense*.

Ventromental plates: W 76 (72-81), L 33 (32-34). VmPR 2.29 (2.21-2.37). Striations (approx.) 25 (22-28). Distance between ventromental plates 47 (41-50) (6). VmPSR 1.66 (1.52-1.97).

Antenna: (Figure 45). L segments 1-5: 51 (49-53), 19 (18-21), 7 (7-8), 9 (8-10), 5 (4-5). AR 0.77 (0.73-0.83). L primary blade 39 (34-42) (5). L secondary blade 9 (9-11) (4). Lauterborn organs reaching beyond middle of segment 3, L Lauterborn organs 6 (5-7) (4). L PS 7 (6-8) (4). ROR 4.77 (3.53-7.39). Segment 1 with 1 sense pit located as in *P. illinoense*. Segment 3 appearing trapezoidal in shape.

Labrum-epipharynx: As in *P. illinoense*. PE with 3-4 teeth in lateral parts, 3 teeth in middle part.

Mandible: L 105 (101-110). L seta subdentalis 16 (16-18), otherwise as in *P. illinoense*.

Maxilla: L lacinia 37 (36-41). L segment 1 maxillary palp 18 (16-20).

Pupa

Description—(n=6, all males) Indistinguishable from *P. illinoense*

Biometrics: L total 3.92 (3.78-4.52) mm L cephalothorax 1.07 (1.00-1.19) mm

Abdomen: Anal taeniae 28.8 (24-32).

Remarks

The species is generally distributed in the eastern U.S., but all the reared specimens are from Florida. Because Florida specimens for most species tend to be smaller than those from other areas of the U.S., the larval and pupal measurements probably represent the small end of the size ranges. Some specimens may be very difficult to distinguish from *P. illinoense* due to the range of variability in *P. illinoense*. The species is named for the sharp bend in the SVo.

Material Examined

—Reared Specimens

Florida, Expressway Creek, Duval Co., W.M. Beck, Jr., 30 Dec. 1964, 4 males, 3 paratypes, 5 May 1965, 1 male paratype, 27 Nov. 1967, 8 males, 3 paratypes; Lake Maitland, Orange Co., 19 May 1964, 1 male; Peter's Creek, Clay Co., 3 Dec. 1964, 7 males, 3 paratypes, 1 April 1966, 3 male paratypes, 15 Nov. 1966, 1 male; Swift Creek, Hamilton Co., 26 Oct. 1965, 1 male; Black Creek, Clay Co., 13 Oct. 1967, 1 male, (Beck). Records from April, May, and October through December.

—Adult Specimens

Illinois, Hardin, Calhoun Co., 10 July 1969, J.C. Martin, 1 male paratype, (INH); Hart Mem. Woods, Mahomet, Champaign Co., 17-18 May 1966, M.W. Sanderson, 2 males, (INH). **Maryland**, Snow Hill, Worcester Co., 1 May 1968, W.H. Anderson, light trap, 3 males, 1 male paratype, 19 May 1968, 1 male, (USNM). **Minnesota**, Red Lake River near Crookston, Polk Co., 17 June 1970, 1 male, 2 July 1970, 1 male paratype, 9 July 1970, 3 male paratypes, 1 male; Lower LaSalle Lake, Hubbard Co., 24 July 1970, 1 paratype, 29 July 1970, 1 male, 1 male paratype; Little Fork, Koochiching Co., 18 June 1968, 1 male, 20 June 1970, 4 males, 3 male paratypes, 21 July 1970, 1 male, 1 male paratype, 28 July 1970, 1 male; Mississippi River at Co. Rd. 2, Clearwater Co., 13 June 1970, 2 males, 5 Aug., 1970, 1 male, 9 Sept 1970, 2 males; all Minnesota specimens from New Jersey mosquito light traps, leg. E.F. Cook (UMn). **New Jersey**, New Brunswick, 11 July 1958, W.W. Wirth, light trap, 1 male paratype, (USNM). **New York**, Cranberry Lake, St. Lawrence Co., 24-26 June 1963, W.W. Wirth, light trap, 1 male paratype, (USNM). **Texas**, 18 mi. W. of Livingston, 24 April 1960, 1 male paratype, (JES). **Virginia**, Falls Church, Holmes Run, 28 June, 1960, 1 male paratype, 17 Oct. 1960, W.W. Wirth, light trap, 1 male, (USNM). Records from April through October.

Polypedilum (*P.*) *angustum* Townes 1945

Polypedilum angustum Townes, 1945: 56; Townes, 1952: 52; Beck and Beck, 1959: 93. Harper and Cloutier, 1979 (biology); Boesel, 1985; 260.

Adult Male

Diagnosis—This species and *P. atrichon* are distinguished from others in *P.* (*Polypedilum*) and *P.* (*Uresipedilum*) by the very narrow Ivo. *P. angustum* can be separated from *P. atrichon* by presence on the former of microtrichia on SVo base, small knob on SVo apex, and spine on foretibial apical scale. SVo distal seta present.

Description—(n=4)

Coloration: Stramineous.

Biometrics: L total 3.76 (3.52-3.93) mm; L thorax 1.01 (0.96-1.06) mm; L wing 1.92 (1.80-2.08) mm; W wing 0.56 (0.53-0.61) mm

Antenna: AR 1.48 (1.43-1.50) (3). D scape 150 (1). L flagellomeres 1-12 460 (440-500) (3). L flagellomere 13 680 (630-750) (3).

Head: W head 510 (510-510) (2). Interocular distance 110 (1). HIR 4.64 (1). Vertical S 14.0 (13-15) (2). Clypeal S 17.5 (14-27). Ocelli not visible. L maxillary palps 1-5: 36 (1), 41 (1), 113 (1), segment 4 not measured, 221 (1).

Thorax: Antepronotum, and thoracic sensilla as in *P. illinoense*. Postpronotal depression with visible campaniform sensilla. Antepronotum without setae. Scutal fossal S absent. Dorsocentral S 11.8 (10-14), uniserial; anterior most seta may be separated from main row by short distance. Prealar S 4.0 (3-5) (2). Scutellar S usually in single row 8.0 (6-11) (3), occasionally 1 or 2 small setae anterior to main row.

Wing: VR 1.16 (1.13-1.21), wing unspotted. R_{4+5} curved downward near wing apex making apex of R_{4+5} closer to wing apex than apex of M. R_1 and R_{2+3} adjacent to distinctly separated apically. Brachiolar S 1. R_1 S 34 (1). R_{4+5} S 23 (22-24) (2). Squamal S 6.7 (6-7).

Legs: L foretibial apical scale 45 (34-56) (3). L spine on foretibial apical scale 10 (9-11) (3), tip may be dark. L mesotibial apical spine 33 (27-38) (2). L hindtibial apical spine 39 (34-45).

	Fe	T1	Ta1	Ta2	Ta3
PI	867 (810-1000)	595 (520-700)	976 (880-1130) (3)	680 (610-780) (3)	466 (400-560) (3)
PII	966 (890-1060) (3)	726 (730-940) (3)	526 (500-580) (3)	270 (250-300) (3)	196 (180-220) (3)
PIII	1070 (960-1200) (3)	900 (790-1010) (3)	725 (670-780) (2)	385 (360-410) (2)	330 (320-340) (2)

	Ta4	Ta5	LR	BV	SV
PI	360 (300-450) (3)	160 (150-180) (3)	1.58 (1.51-1.63) (3)	1.49 (1.44-1.53) (3)	1.54 (1.49-1.58) (3)
PII	133 (120-150) (3)	68 (60-75) (3)	0.64 (0.62-0.69) (3)	3.48 (3.37-3.56) (3)	3.41 (3.24-3.52) (3)
PIII	210 (1)	100 (1)	0.76 (0.74-0.77) (2)	2.66 (1)	2.89 (2.86-2.92) (2)

Hypopygium: (n=5, including holotype) (Figure 46). Tergite IX S 10.0 (8-12) (4), located in broad middorsal area. Anal point about even with apices of IVo, -5 (0 to -18). SVo usually narrow, curved slightly medially, and with small apical knob (apical knob not shown in Figure 46). L SVo 81 (72-92). Placement of distal seta variable; mean SVoR 2.49 (1.56-3.62) (5). Base of SVo with weak, but relatively numerous, microtrichia. Basal S 3.4 (2-5). Max L distal S 70 (63-76), not exceeding IVo in length. IVo long and very narrow, sides subparallel, surface covered with microtrichia except dorsally in area bearing dorsal setae. L IVo 117 (107-126). W IVo 20 (17-23). IVo dorsal setae 9.9 (7-13), moderately developed; posteriorly directed apical seta usually present. Gonostylar medial margin nearly straight, lateral margin evenly convex. Gonostylus of holotype with apical notch. Medial S 7.4 (5-8), weak, not reaching midline. Max L medial S 57 (52-68) (4). Max L apical S 36 (36-36) (2). L gonostylus 179 (158-203). GsR 3.40 (3.08-3.82). HR 2.78 (2.33-3.08). L posterior lobe gonocoxite 65 (54-78). HV 2.11 (1.96-2.32) (3). Microtrichia on basal plate numerous but relatively small. Phallapodeme usually straight and narrow.

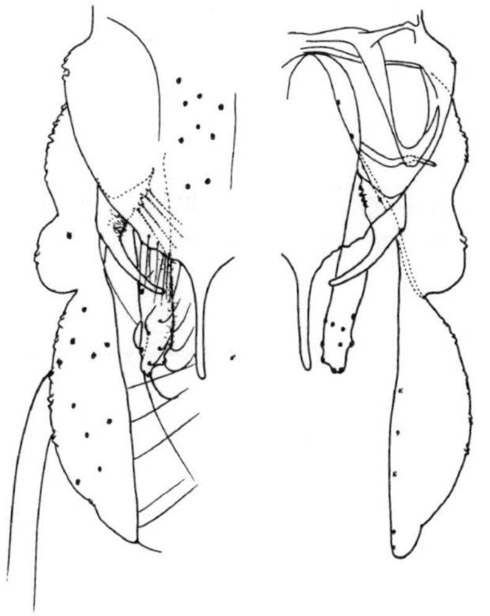

Figure 46. *P. angustum* Townes, hypopygium.

Type locality: New Jersey, bank of Rancocas Creek, Riverside, 18 June 1939, male, H.K. Townes, (HKT). *Polypedilum angustum* seems to include several potential subgroups. One of these is *P.* (*P.*) *atrichon*, sp. nov., but other new species may be included in a *P.* (*P.*) *angustum* complex.

Biology

Harper and Cloutier (1979) found *P. angustum* in small numbers throughout the summer, associated with the slower moving reaches of the L'Achigan River, Quebec.

Material Examined—Adult Specimens

Holotype male (slide of male hypopygium). **Minnesota**, Cedar Creek Natural History Area, Isanti Co., Beckman Lake, June, July 1971, emergence trap over acid bog, 7 males, Roberta Lammers. **Virginia**, Alexandria, Dyke Marsh, 13 May 1958, 1 male, W.W. Wirth, (USNM). **Massachusetts**, Bedford, 21 July 1961, 1 male, W.W. Wirth, (USNM); **Illinois**, Brussels, 26 Sept. 1939, Frison and Ross, 1 male, (INH). Records from May through July and September.

Polypedilum (*P.*) *artifer* (Curran) 1930

Chironomus artifer Curran, 1930: 32.

Polypedilum artifer (Curran), Townes, 1945: 53; Hauber, 1947: 462; Sublette, 1960: 207; Sublette, 1966: 26, 29 (description of holotype including figure of hypopygium); Frommer, 1967: Figure 120 (figure of anterior portion of thorax); Boesel, 1985: 258.

Adult Male

Diagnosis—Forefemur stramineous, except shaded slightly darker toward proximal end; otherwise as in *P. fallax*.

Description—(n=8)

Coloration: As in *P. fallax* except as noted in diagnosis.

Biometrics: L total 5.51 (4.10-6.07) mm; L thorax 1.48 (1.07-1.72) mm; L wing 2.97 (2.25-3.43) mm; W wing 0.85 (0.63-0.96) mm

Antenna: AR 1.20 (1.09-1.37) (7). D scape 206 (170-220) (7). L pedicel 143 (130-150) (3). L flagellomeres 1-12 681 (570-750) (7). L flagellomere 13 813 (630-910) (7).

Head: W 661 (520-750). Interocular distance 100 (70-120). HIR 6.61 (5.58-7.43). Vertical S 19.1 (12-24) (7). Clypeal S 23.3 (17-32). Ocelli as in *P. fallax*. L maxillary palp segments 1-5: 53 (41-59), 57 (50-61), 196 (140-230), 173 (119-209), 287 (185-374).

Thorax: As in *P. fallax*. Acrostichal S 21 (1). Dorsocentral S 19.7 (12-28). Scutal fossal S 2.9 (1-4). Prealar S 10.4 (6-14). Scutellar S 28.4 (15-39), in 2 to 3 rows.

Wing: VR 1.18 (1.15-1.23). Brachiolar S 1.1 (1-2). R_1 S 59 (49-63). R_{4+5} S 55 (38-73). Squamal S (approx.) 30 (14-36).

Legs: Foretibial apical scale as in *P. fallax*, total L 73 (56-90). L mesotibial apical spine 31 (25-34). L hindtibial apical spine 31 (28-36) (7).

	Fe	T1	Ta1	Ta2	Ta3
PI	1335 (1000-1560)	1023 (760-1150)	1558 (1130-1770)	1043 (790-1160)	801 (590-900)
PII	1565 (1170-1880)	1301 (970-1540)	758 (550-900)	493 (350-590)	384 (270-450)

PIII	1691 (1250-1990)	1464 (1100-1690)	1121 (810-1330)	645 (470-770)	530 (390-620)
	Ta4	Ta5	LR	BV	SV
PI	645 (490-740)	260 (160-310)	1.52 (1.47-1.60)	1.43 (1.36-1.53)	1.52 (1.45-1.57)
PII	244 (180-290)	131 (100-160)	0.58 (0.57-0.60)	2.92 (2.80-3.01)	3.81 (3.61-3.92)
PIII	318 (240-350)	154 (120-180)	0.77 (0.73-0.79)	2.61 (2.46-2.69)	2.81 (2.62-2.93)

Hypopygium: (n=9, including the holotype) As in *P. fallax* except anal point generally shorter in relation to apices of IVo, -26 (7 to -61). Tergite IX S 11.0 (6-14). SVo basal S 3.4 (2-5). Mean SVoR 0.78 (0.59-0.91). Max L distal S 131 (117-158). L IVo 193 (153-250). W IVo 41 (27-61). IVo dorsal S 24.9 (14-34). L gonostylus 237 (162-270). GsR 3.95 (3.28-5.22). Gonostylar medial S 5.8 (4-8) (8), in irregular row in some specimens. Max L medial S 100 (72-117). Max L apical S 54 (45-59) (5). L posterior lobe gonocoxite 104 (86-128). HR 2.30 (1.90-3.00). HV 2.35 (2.18-2.58) (8).

Type locality: New York, Tuxedo Park, 1 July, (year?) male, (AMNH).

Material Examined — Adult Specimens

Holotype male. **Maine**, Lake Pushaw, Old Town, 1 Aug. 1966, W.W. Wirth, at light, 1 male, (USNM). **Minnesota**, Lower LaSalle Lake, Hubbard Co., 12 Aug., 8 Sept., 12 Sept. 1970 leg. E.F. Cook, New Jersey mosquito light trap, 3 males, (UMn); U.S. EPA Lab., Duluth, St. Louis Co., 4 Aug. 1970, leg. E.F. Cook, New Jersey mosquito light trap, 1 male, (UMn); Minnesota Forestry Station, Little Fork, Koochiching Co., 15 July, 26 Sept. 1970, leg. E.F. Cook, New Jersey mosquito light trap, 2 males, (UMn); Coon Creek, Coon Rapids, Anoka Co., 45°08'N 93°18'W, 7 Aug. 1969, D.E. Maschwitz, 2 males, (UMn). **New Hampshire**, Stinson Lake, White Mountains, 23 July 1961, W.W. Wirth, 1 male, (USNM). **New York**, Genesee R., Portageville, 13 June 1963, W.W. Wirth, 1 male, (USNM). **Washington**, Silver Springs Campground, 31 mi. SE Enumclaw, 15 Sept. 1967, D. Hansen, at illuminated sheet by springs, 1 male, (UMn); from June through September.

Polypedilum (P.) atrichon New Species

Adult Male

Diagnosis—Similar to *P. angustum* but distinguished from that species by the following: Base of SVo small, without microtrichia, SVo broad near base, tapered to blunt point at apex. Foretibial apical scale without spine (see Remarks).

Description—(n=5)

Coloration: Stramineous. Pedicel and flagellum of antenna light brown, thorax occasionally tinged with light brown.

Biometrics:	L total 4.17 (3.85-4.39) mm	L thorax 1.07 (1.04-1.11) mm
	L wing 2.13 (1.99-2.20) mm	W wing 0.61 (0.56-0.68) mm

Antenna: AR 1.46 (1.42-1.50) (4). D scape 153 (140-160) (3). L pedicel 120 (120-120) (4). L flagellomeres 1-12 492 (480-510) (4). L flagellomere 13 (720 (690-760) (4).

Head: W head 507 (480-520) (4). Interocular distance 107 (100-110) (4). HIR 4.73 (4.36-5.10) (4). Vertical S 13.9 (11-16). Clypeal S 17.4 (15-20). Ocelli very small, not visible on some specimens. L maxillary palps 1-5: 40 (36-45), 42 (41- 45), 133 (108-146), 136 (131-144), 202 (178-225).

Thorax: Antepronotum similar to *P. illinoense*. Postpronotal depression with campaniform sensilla distinct. Humeral pit present in some specimens. Antepronotum without setae. Dorsocentral setae 12.2 (9-15). Scutal fossal S absent. Prealar S 4.9 (4-6). Scutellar S 9.8 (7-14) in single row of large setae except in 1 specimen with anterior row of small setae.

Wing: VR 1.16 (1.14-1.18). Wing unspotted. Apices of R_{4+5} and M about equidistant from wing apex. Apex of R_{4+5} curving moderately downward in some specimens. R_1 and R_{2+3} adjacent to distinctly separated apically. Brachiolar S 1. R_1 S 50.5 (43-63) (4). R_{4+5} S 37.0 (32-44) (4). Squamal S 8.7 (6-11) (4).

Legs: Foretibial apical scale without apical spine. L scale 43 (38-47). L mesotibial apical spine 26 (23-29). L hindtibial apical spine 24 (23-29).

	Fe	**T1**	**Ta1**	**Ta2**	**Ta3**
PI	904 (860-950)	658 (600-710)	1108 (1040-1160)	776 (760-790) (3)	536 (530-540) (3)
PII	1026 (1020-1030) (3)	903 (890-930) (3)	567 (560-590) (3)	313 (300-330) (3)	226 (220-240) (3)
PIII	1056 (1050-1070) (3)	1006 (1000-1020) (3)	790 (770-810) (2)	435 (430-440) (2)	345 (340-350) (2)

	Ta4	**Ta5**	**LR**	**BV**	**SV**
PI	393 (390-400) (3)	153 (150-160) (3)	1.70 (1.60-1.81)	1.47 (1.43-1.51) (3)	1.41 (1.35-1.45) (3)
PII	143 (130-150) (3)	83 (80-90) (3)	0.64 (0.62-0.66) (3)	3.29 (3.18-3.41) (3)	3.37 (3.27-3.45) (3)
PIII	220 (210-230) (2)	100 (90-110) (2)	0.78 (0.76-0.80) (2)	2.61 (2.57-2.65) (2)	2.64 (2.53-2.69) (2)

Hypopygium: (Figure 47). Tergite IX S 6.0 (2-8), located in middorsal triangular shaped area. Anal point not reaching apex of IVo, -28 (-11 to -36). SVo broad proximally, evenly tapered to blunt point apically. SVo usually evenly curved medially. Base of SVo very small, without microtrichia. Basal S 2.1 (2-3). LSVo 87 (86-88) (1). Distal seta absent except on one specimen; SVoR 2.23 (2.00-2.45) (1). Max L distal S 70 (1), not exceeding apex of IVo. IVo as in *P. angustum*. L IVo 133 (127-145). W IVo 25 (23-27). IVo dorsal S 14.9 (12-18); posteriorly directed apical seta present. Gonostylus as in *P. angustum*. Medial S 4.2 (3-5), weak, not reaching midline. Max L medial S 47 (36-61). Max L apical S 31 (23-41). L gonostylus 176 (158-189). GsR 3.83 (3.50-4.26). HR 2.63 (2.12-3.00). L posterior lobe gonocoxite 67 (61-74). HV 2.38 (2.20-2.58). Phallapodeme straight, generally narrow, expanded at medial end.

Type locality: Holotype male, **Minnesota**, marshy pond, Shoreview, Ramsey Co , 45°05'N 93°08'W, 18 July 1969, D.E. Maschwitz.

Remarks

The holotype male was obtained from a mass-rearing from leaf petioles and main stems of *Potamogeton natans*, (UMn). The remaining specimens examined are designated as paratypes.

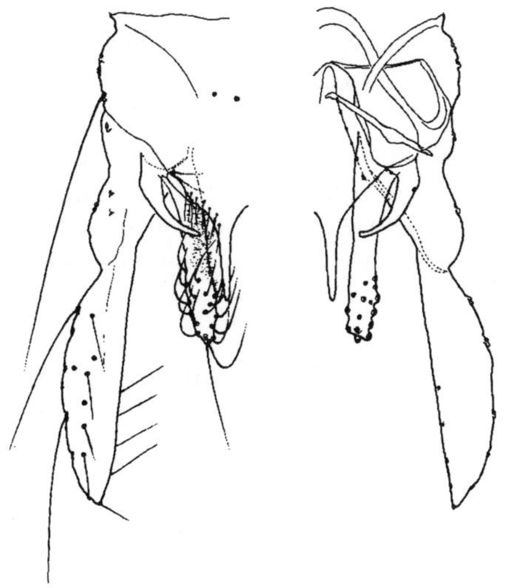

Figure 47. *P. atrichon* sp. nov., hypopygium.

A single male specimen (Ontario, White Lake, 25 May 1968) is essentially the same as the other specimens except for slight hypopygial differences and the presence of the apical spine on the foretibial apical scale. This specimen is included as this species. *Polypedilum atrichon* is named for the fact that the base of the SVo lacks microtrichia and the SVo distal seta is usually absent.

Biology
This larvae of this species were found on the stems and leaf petioles of *Potamogeton natans*. They probably live in elongate cells cut into the tissue along the long axis of the stems.

Material Examined—Adult Specimens
Minnesota, marshy pond, Shoreview, Ramsey Co , 45°05'N 93°08'W, 18 July 1969, D.E. Maschwitz, 3 males (UMn). **Ontario**, Costello Lake, Algonquin Park, 21 Aug. 1964, D.W. Webb, 1 male (INH); White Lake, 25 May 1968, R.D. MacDonald, 1 male, (CNC). Records from May, July, and August.

Polypedilum (U.) aviceps Townes 1945

Polypedilum aviceps Townes, 1945: 61; Roback, 1953: 120 (sp. "a" probably is *P. aviceps*, biology); Sublette, 1960: 208; Beck, 1977 (biology); Harper and Cloutier, 1979 (biology); Boesel, 1985; Soponis and Russell, 1982 (characters of early instar larvae); Soponis, 1983 (emergence patterns); Niitsuma, 1992 (description, figure of hypopygium); Sasa and Kikuchi, 1995 (figure of hypopygium); Oyewo and Sæther (1998).

Adult Male

Diagnosis—Anal point very broad. Medial process near apex of SVo small compared to other *P.* (*Uresipedilum*) (*convictum* group) species. IVo small, subparallel sided, rounded apically.

Description—(n=7)

Coloration: Head stramineous to light brown, pedicel and flagellum light brown, always darker than head. Thoracic vittae, postnotum, and preepisternum, especially ventral part, light brown. Haltere stramineous. Legs stramineous to light brown with foretarsi shade darker. Abdomen and hypopygium stramineous. Few specimens are entirely stramineous.

Biometrics: L total 3.92 (3.47-4.53) mm; L thorax 1.10 (0.99-1.21) mm; L wing 2.22 (1.88-2.45) mm; W wing 0.68 (0.60-0.73) mm

Antenna: AR 1.66 (1.50-1.95) (6). D scape 162 (150-180). L pedicel 125 (120-130) (2). L flagellomeres 1-12 433 (420-450) (6). L flagellomere 13 723 (630-880) (6).

Head: W head 516 (490-560). Interocular distance 112 (90-140). HIR 4.63 (4.00-5.44). Vertical S 13.2 (10-16) (6). Clypeal S 15.0 (11-21). Ocelli very small, not seen in many specimens. L maxillary palp 1-5: 42 (38-45) (5), 33 (27-41) (5), 185 (149-276) (5), 144 (133-164) (6), 232-(207-266) (5).

Thorax: Prothorax similar to *P. illinoense* but reduced; antepronotum not extending as far dorsally as in other *Polypedilum* species. Postpronotal depression small, campaniform sensilla not visible. Humeral pit present. Dorsocentral S 14.4 (10-21) (6), uniserial. Prealar S 5.0 (4-7) (6). Scutellar S in 2 rows, setae in anterior row nearly as large as setae in posterior row; anterior row S 7.5 (6-9) (6), posterior row S 11.1 (8-13) (6).

Wing: VR 1.20 (1.17-1.22). R_{4+5} and M variable in position apically, usually about equidistant from wing apex. R_1 and R_{2+3} adjacent apically. Brachiolar S 1. R_1 S 37.8 (33-47) (6). R_{4+5} S 28.3 (23-34). Squamal S 12.2 (10-17).

Legs: Foretibial apical scale without spine. L apical scale 34 (32-38) (5). L mesotibial apical spine 27 (23-32). L hindtibial apical spine 28 (20-32).

	Fe	T1	Ta1	Ta2	Ta3
PI	904 (850-1000)	700 (640-760)	1024 (970-1070)	738 (700-810)	525 (490-580)
PII	1042 (980-1140)	891 (820-980)	525 (480-570)	291 (260-320)	224 (200-250)
PIII	1075 (1000-1200)	1012 (910-1100)	768 (700-830)	404 (370-450)	340 (310-380)

	Ta4	Ta5	LR	BV	SV
PI	405 (380-440)	170 (150-200)	1.47 (1.41-1.52)	1.43 (1.39-1.46)	1.55 (1.49-1.62)
PII	142 (130-160)	87 (70-100)	0.58 (0.56-0.62)	3.31 (3.15-3.52)	3.69 (3.54-3.85)
PIII	205 (190-220)	104 (90-120)	0.74 (0.66-0.77)	2.71 (2.66-2.85)	2.72 (2.64-2.79)

Hypopygium: (n=8) (Figure 48). Tergite IX S 19.5 (13-32) (8), in broad middorsal oval shaped area. Anal point curved ventrally, dorsal margin broadly expanded laterally. Anal point usually exceeding IVo, 23 (-11 to 43). SVo broad, medial process digitiform, smaller than in other *P.* (*Uresipedilum*) (*convictum* group) species. No tubercle at origin of distal seta. Microtrichia numerous on apical portion of SVo. Basal S 1.1 (1-2) (8) (1 specimen had 2 seta on 1 SVo), positioned as in *P. flavum*. LSVo 98 (83-113). Mean SVoR 1.10 (0.88-1.29) (8). Max L distal S 76 (61-90) (8). IVo small compared to other species in *P.* (*Uresipedilum*), medial and lateral margins subparallel, or IVo expanded slightly apically; apex rounded. Ventral, subapical, posteriorly directed seta present. Dorsal surface along lateral margin free of microtrichia. L IVo 90 (83-96) (8). W IVo 25 (23-29) (7). IVo dorsal S 16.3 (14-21) (5), weakly developed. Gonostylus similar to *P. flavum*, medial margin often slightly sinuate, straight in some specimens. Medial S 8.3 (7-10) (8), weakly developed. Max L medial S 37 (34-45) (8). Max L apical S 30 (25-34) (8), usually nearly as long as medial setae. L gonostylus 167 (142-191) (8). GsR 3.20 (2.83-3.50) (8). HR 3.15 (2.51-3.86) (8). L posterior lobe gonocoxite 53 (45-68) (8). HV 2.32 (1.83-2.54) (7). Phallapodeme usually moderately expanded midway along its length and near medial end.

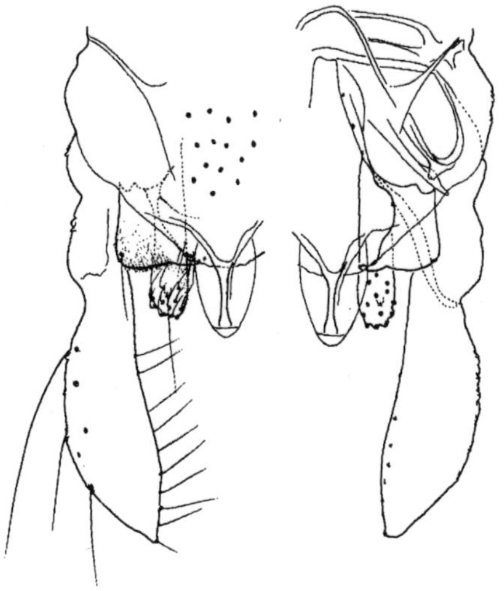

Figure 48. *P. aviceps*, hypopygium.

Type locality: New York, Ithaca, 11 Sept. 1935, H.K. Townes, male, (HKT).

Larva

Diagnosis—Second lateral teeth relatively large, more than half as long as median 2 teeth. Antennal segment 3 small, AR 0.57.

Description—(n=4, 3 males, one unknown)

Biometrics: L body 5.4 (1) mm L head capsule 192 (187-198) (3)

Mentum: (Figure 49). W 105 (92-115) (3), with 16 teeth, last lateral teeth very small, median 2 and second lateral teeth large. First lateral teeth nearly as large as second lateral teeth, reaching beyond the mid point of the median 2 teeth. Fourth through seventh lateral teeth smaller, becoming progressively smaller laterally. Teeth uniformly dark, bluntly pointed.

Figure 49. *P. aviceps* Townes, mentum.

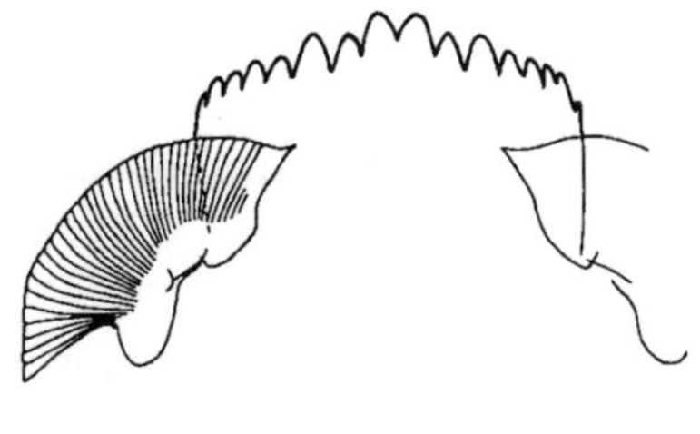

Ventromental plates: W 85 (81-88). L 43 (41-47). VmPR 1.98 (1.85-2.08) (3). Striations (approx.) 33, distinct. Posterior lobe well developed. Distance between ventromental plates 53 (50-56) (3). VmPSR 1.64 (1.54-1.73) (3).

Antenna: (Figure 50). L segments 1-5: 85 (79-91), 32 (30-33), 5 (3-6), 7 (7-8), 5 (5-5). AR 0.57 (0.57-0.58). L primary blade 39 (1), apparently does not exceed apex of segment 5. L secondary blade 11 (10-11) (2). Lauterborn organs small, reaching just to apex of segment 3, very transparent and difficult to see in most specimens. L Lauterborn organs 6 (1). Triangular structures not visible. L PS 6 (5-6) (2), lanceolate. ROR 4.91 (4.00-5.79) (3). One sense pit located near apex of segment 1.

Labrum-epipharynx: Labral S I as in *P. illinoense*, S II narrow, with filamentous branches on both sides, S III and IV as in *P. nubeculosum*. PE with 5-6 teeth in lateral 2 parts, mid part with 3-4 teeth.

Mandible: L 116 (110-122) (3), 5 teeth, point of fifth tooth definitely free from, and projecting beyond, medial margin. Medial margin with 3 needle-like serrations. L seta subdentalis 20 (18-20) (3). Seta interna with 4 main branches apparently with common origin. PMa with about 10-12 filaments.

Maxilla: Lacinia as in *P. flavum*, L 44 (36-58). Galea and maxillary palp as in *P. flavum*. L segment 1 maxillary palp 19 (18-20).

Pupa

Diagnosis—This species distinguished from *P. flavum* and *P. obtusum* by small number of anal taeniae and presence of pedes spurii A.

Description—(n=3, all males)

Biometrics: L total 3.61 (3.40-3.89) mm   L cephalothorax 1.07 (0.98-1.19) mm

Coloration: Typical.

Abdomen: (Figure 51). Pedes spurii B well developed on segments I and II; pedes spurii A present. Hooks on tergite II larger than spinules but light in color and not prominent. Spinules on tergite II as in *P. flavum*. Spinule pattern on tergite III through VIII similar to *P. obtusum* but reduced. Intersegmental spinules as in *P. flavum*. Lateral taeniae moderately well developed, 3 on segments V and VI; 4 on segments VII and VIII. Anal spur with 1 primary point and 0 to 3 small apical accessory points. Anal taeniae 19.0 (18-20) on each anal lobe.

50

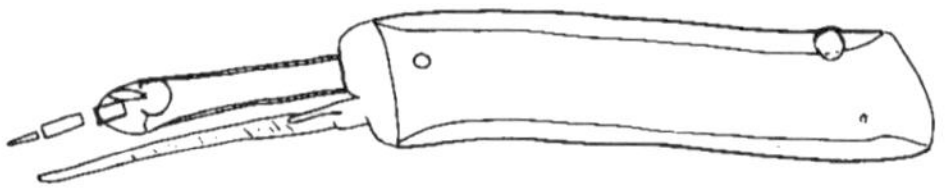

51

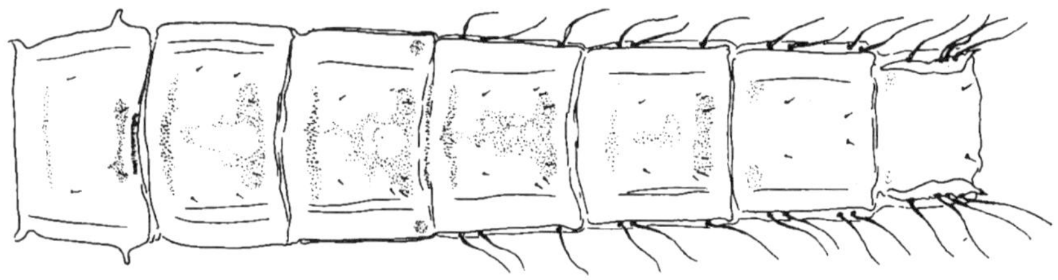

Figures 50-51. *P. aviceps* Townes: 50-antenna; 51-pupal abdomen.

Biology

Polypedilum aviceps is a fairly common, widely distributed species. We reared specimens from larvae collected in algae and plant debris caught in branches overhanging a trout stream (Sucker Brook). We saw a single specimen of Roback's (1953) species "a" and have confirmed that it is *P. aviceps*. Roback found this species on submerged wood and on plants in swift water. *Polypedilum aviceps* is a very common species in the L'Achigan River, Quebec, where it shares the riffles with *P. convictum* (*P. flavum* ?). In an apparent competitive separation, the former species prefers the lower sections and the latter the upper sections of riffles (Harper and Cloutier, 1979). *Polypedilum aviceps* has a distinct bimodal emergence pattern in the L'Achigan River; a short but intense first emergence around the first of June and a second in mid-July (Harper and Cloutier, 1979). *Polypedilum aviceps* also has a bimodal emergence pattern in Turkey Creek, northern Florida; but, at this latitude (about 30° N), the first emergence is in February/March and the second in November (Soponis, 1983). *Polypedilum aviceps* may be less tolerant of pollution than *P. convictum* (*P. flavum* ?) (Hudson *et al.*, 1990).

Material Examined

—Reared Specimens

Florida, Canoe Creek, Escambia Co., 20 Oct. 1967, W.M. Beck, Jr., 3 males, (Beck); Black Creek, Clay Co., 23 Aug. 1966, W.M. Beck, Jr., 1 male (Beck); Swift Creek, Hamilton Co., 19 March 1968, W.M. Beck, Jr., 1 male (Beck). **Minnesota**, Sucker Brook, Clearwater Co., 47°16'N 95°16'W, 24 Aug. 1969, D.E. Maschwitz, 1 male, (UMn). Records from March, August,and October.

—Adult Specimens

Paratype males (slides of hypopygium) no data, 4 males (USNM). **Illinois**, Dundee, Kane Co., 23 May 1939, Burke and Riegel, 1 male, (INH); New Delhi, Jersey Co., 10 July 1969, J.C. Martin, 1 male, (INH). **Minnesota**, numerous specimens collected in New Jersey mosquito light traps in June, July, and Aug. from Hovland, Cook Co.; Duluth, St. Louis Co.; Mississippi R. at Co. Rd. 2, Clearwater Co.; Lower LaSalle Lake, Hubbard Co., (UMn); Minneapolis, at light, 29 May 1969, D.E. Maschwitz, 1 male, (UMn); Sucker Brook, Clearwater Co., 47°16'N 95°16'W, mass reared from algae and debris in overhanging branches, 24 Aug. 1969, D.E. Maschwitz, 1 male, (UMn). **Montana**, Hamilton, 13 Aug. 1965, malaise trap, 2 males (JES). **New Hampshire**, Stinson Lake, White Mts., 23 July 1961, W.W. Wirth, 1 male (USNM). **New Jersey**, Elizabeth Town, 3 July 1964, Goodwin, 1 male, (JES). **New York**, several localities, June, 6 males (USNM). **Oregon**, Salmon R. Hwy 18, Rose Lodge, 13 Aug. 1968, J.E. Sublette, at light, 14 males (JES); Willamette R., Salem, 22 May 1963, K. Goeden, light, 1 male, (USNM); Aumsville, June 1963, D. Goeden, light trap, 1 male, (USNM). **Pennsylvania**, Conewago Creek, 13 mi. W York, 19-21 June 1959, J.P. Spangler, at light, 1 male, (USNM). **Utah**, Hwy. Bridge 89-91, Provo R. 2 July 1947, G.F. Edwards, 5 males (JES). **Virginia**, Vesuvius, Rockbridge Co., July 1960, D.H. Messersmith, light trap, 31 males (USNM). **Wisconsin**, 11 mi. E, 4 mi. S Siren, Burnett Co., 16 July 1966, D. Hansen, light trap, 1 male, (UMn). **Wyoming**, S Fork Clear Creek, S Fork Campground, 16 mi. WSW Buffalo, 23 Aug. 1967, attracted to illuminated sheet, D. Hansen, 5 males (UMn). **New Brunswick**, NW Miramichi R., Northumb Co., 22 July 1966, J. Marshall, 2 males, (CNC). **Ontario**, Niagara Falls, Welland Co., 19 Sept. 1967, D.W. Webb, 1 male (INH). Records from May through September.

Polypedilum (*P.*) *bergi* New Species

Adult Male

Diagnosis—Antennal ratio more than 1.60. SVo curved medially as in *P. falciforme* but usually heavier and thicker than in latter species. Dorsal setae usually concentrated toward the posterolateral margin of IVo. Usually more than 4 medial setae on gonostylus.

Description—(n=9)

Coloration: Head, thorax, abdomen, wing veins, haltere, and legs uniformly stramineous. Pedicel tinged with light brown in some specimens. Flagellum light brown.

Biometrics:	L total 4.43 (3.80-5.26) mm	L thorax 1.19 (1.04-1.31) mm
	L wing 2.36 (2.17-2.61) mm	W wing 0.69 (0.63-0.74) mm

Antenna: AR 1.78 (1.65-1.90). D scape 178 (170-200) (8). L pedicel 120 (110-130) (3). L flagellomeres 1-12 477 (430-520). L flagellomere 13 848 (760-950).

Head: W 610 (540-660) (8). Interocular distance 113 (80-130) (8). HIR 5.40 (4.77-7.25) (8). Vertical S 15.3 (12-20) (8). Clypeal S 19.2 (15-22) (8). Ocelli small, not visible in some specimens. L maxillary palp segments 1-5: 45 (38-56) (7), 49 (38-56) (7), 49 (41-54) (7), 129 (115-142) (7), 125 (119-128) (7), 221 (200-250) (7).

Thorax: Antepronotum as in *P. illinoense*. Antepronotal and scutal sensilla similar to *P. illinoense*. Antepronotum without setae. Acrostichal S 16 (1). Scutal fossal S 2 (0-2) (present on two of eight measured specimens, absent on others). Dorsocentral S 16.6 (12-23) (8), uniserial. Prealar S 5.7 (5-7) (8). Scutellar S in 2 regular or irregular rows, setae in anterior row smaller; anterior row 10.3 (7-12) (8); posterior row 12.5 (9-17) (8).

Wing: VR 1.15 (1.12-1.18) (8), wing unspotted. Apices of R_{4+5} and M either equidistant from wing apex (4 of 8 specimens) or R_{4+5} closer to apex than M. R_1 and R_{2+3} adjacent or slightly separated apically. Brachiolar S 1 (7). R_1 S 50.1 (41-58) (8). R_{4+5} S 42.4 (30-50) (7). Squamal S 13.3 (11-16) (8).

Legs: L foretibial apical scale 44 (34-50) (8). L spine on foretibial apical scale 11 (7-14) (8), darkened at tip in some specimens. L mesotibial apical spine 34 (25-36) (5). L hindtibial apical spine 37 (32-41) (6).

	Fe	T1	Ta1	Ta2	Ta3
PI	1054 (930-1170)	733 (660-820)	1253 (1150-1440)	872 (800-970)	646 (580-730)
PII	1187 (1040-1330) (8)	1001 (900-1120) (8)	631 (540-770) (7)	382 (330-450) (7)	292 (250-340) (7)
PIII	1267 (1100-1370) (8)	1185 (1050-1320) (8)	906 (790-1030) (8)	516 (440-600) (8)	428 (360-500) (8)

	Ta4	Ta5	LR	BV	SV
PI	490 (430-550)	191 (170-220)	1.73 (1.58-1.93)	1.39 (1.36-1.46)	1.42 (1.32-1.55)
PII	174 (150-210) (7)	95 (80-110) (7)	0.63 (0.59-0.68) (7)	3.02 (2.87-3.22) (7)	3.52 (3.21-3.73) (7)
PIII	251 (210-300) (8)	115 (100-130) (8)	0.76 (0.72-0.78) (8)	2.58 (2.45-2.73) (8)	2.73 (2.61-2.88) (8)

Hypopygium: (n=10) (Figure 52). Tergite IX S 10.8 (6-24), usually located in long narrow middorsal area, occasionally broader anteriorly or indistinctly demarcated in some specimens. Anal point straight or curved ventrally. L anal point variable in relation to apex of IVo, -5 (23 to -41). SVo similar to *P. falciforme* but more variable. SVo curved medially, apex pointing toward area near base of anal point, approximately perpendicular to long axis of anal point. Some specimens with as much as apical 1/2 of SVo straight; in others, apex curved so that apex points more anteriorly on anal point. SVo narrow, sickle-shaped to large, thick, and heavy, usually heavier than in *P. falciforme*. Basal S 3.0 (2-4). Microtrichia on SVo base numerous. LSVo 112 (97-126). Placement of distal S variable. Mean SVoR 1.90 (1.26-2.91). Max L distal S 100 (68-119) (9), moderately weak, usually just reaching apex of IVo, exceeding it in few specimens. IVo medial margin usually straight (convex in some specimens), subparallel to lateral margin. L IVo 133 (101-158). W IVo 52 (45-68) (9). Dorsal S 17.7 (12-22), moderately well developed; usually concentrated on lateral margin. IVo usually with 1 to 3 small posterior setate papillae. Gonostylus as in *P. illinoense*. Medial S 5.4 (4-7) (8), moderately weak. Max L medial S 60 (41-70) (8). Max L apical S 24 (14-32) (8). L gonostylus 203 (164-248). GsR 4.72 (3.90-5.39) (9). HR 3.49 (2.93-3.89). L posterior lobe gonocoxite 58 (50-68). HV 2.27 (2.07-2.47) (9). Phallapodeme as in *P. falciforme*.

Type locality: Holotype male, **Minnesota**, marshy pond, Shoreview, Ramsey Co., 45°05'N 93°08'W, 10 Aug. 1971, D.E. Maschwitz, (UMn).

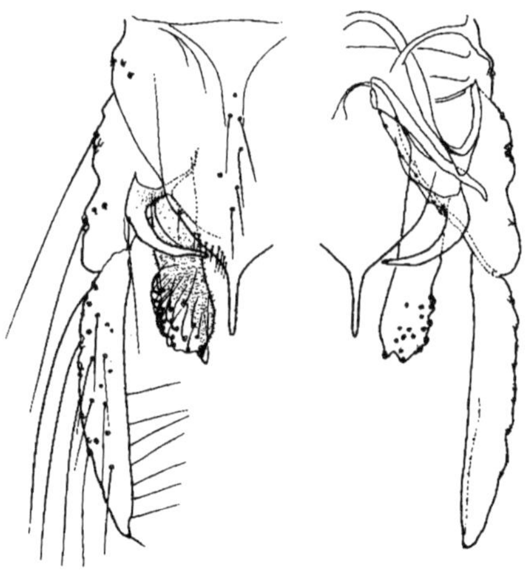

Figure 52. *P. bergi*, sp. nov., hypopygium.

Larva

Diagnosis—Median 2 teeth of mentum distinctly smaller than second lateral teeth, first lateral teeth very small.

Description—(n=5; 3 males, 2 females)

Live color: pale red or orange.

Biometrics: L total 6.83 (6.21-7.70) (4) mm L head capsule 235 (214-263)

Mentum: (Figure 53). W 112 (95-126), with 16 teeth, second lateral teeth large, first lateral and last lateral teeth very small and subequal in size. Median 2 teeth about the same size or slightly smaller than lateral teeth 3 through 6. Teeth subtriangular in shape, generally show little wear.

Ventromental plates: W 95 (88-106). L 43 (38-45), posterior lobe weakly developed. VmPR 2.21 (1.95-2.41). Striations (approx.) 33, distinct. Distance between ventromental plates 54 (50-61). VmPSR 1.77 (1.63-2.00).

Antenna: (Figure 54). L segments 1-5: 62 (52-69) (4), 21 (20-22) (4), 10 (9-11) (4), 10 (10-11) (4), 4 (3-5) (4). AR 0.72 (0.67-0.83) (4). L primary blade 36 (32-39) (3), not exceeding apex of segment 5. L secondary blade 11 (10-12) (2). L PS 10 (10-11) (3). L Lauterborn organ 7 (7-7) (2), arch shaped, reaching middle of segment 3. Triangular structures at base of Lauterborn organs visible. ROR 4.23 (4.06-4.53) (3). Segment 1 with 1 sense pit near apex.

Labrum-epipharynx: As in *P. illinoense* except labral S II narrow with distinct filamentous branching, and PE in three parts, middle part with 1 or 2 teeth, lateral parts with 2 to 3 teeth.

Mandible: L 137 (128-146), as in *P. illinoense*, except as noted: medial margin usually with 3 serrations, some specimens with 2. L seta subdentalis 22 (18-25). PMa with 6 (5-7) long filaments.

Maxilla: Lacinia long, beak-like in appearance, L lacinia 41 (36-45). Galea slightly to moderately developed. L segment 1 maxillary palp 23 (20-25) (4), segments 2 and 3 small.

53

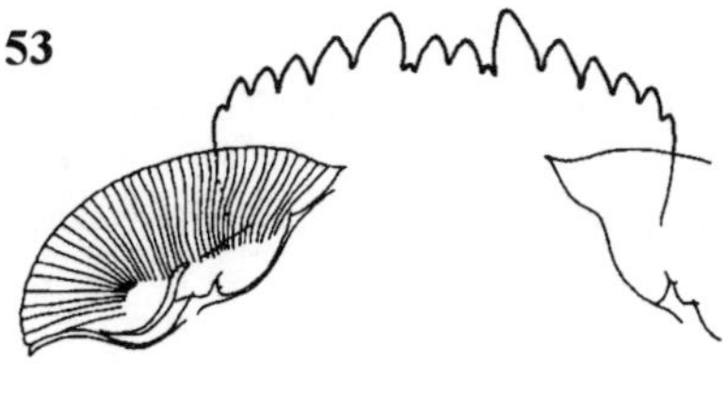

54

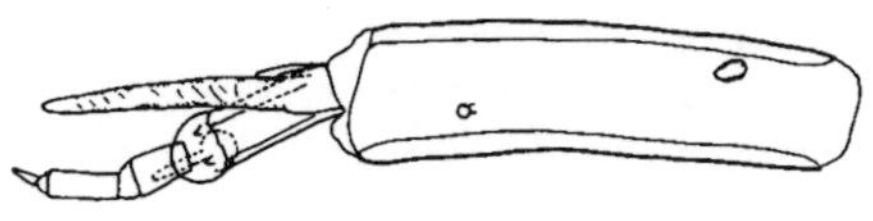

55

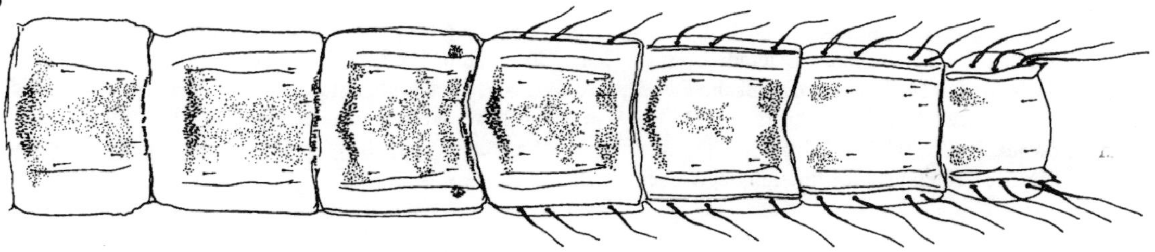

Figures 53-55. *P. bergi* sp. nov.: 53-mentum; 54-antenna; 55-pupal abdomen.

Pupa

Diagnosis—Very close to *P. falciforme* but generally larger than this species and with more numerous anal taeniae, (range: 44-67). Cephalothorax with prominent tubercle anterior to wing sheath (see Figure 74, page 69).

Description—(n-4; 3 males, 1 female)

Biometrics: L total 5.87 (5.47-6.21) mm L cephalothorax 1.34 (1.26-1.49) mm

Coloration: Typical.

Abdomen: (Figure 55). Pedes spurii B on segments I and II, and pedes spurii A on segment IV well developed. Hooks on tergite II generally smaller than large anterior spinules on tergites III-V. Intersegmental spinules between tergites III and IV uniserial to irregularly biserial; those between tergites IV and V biserial to multiserial. Intersegmental spinules pointed anteriorly, generally larger than anterior spinules. Shagreen spinules small, forming a triangular shaped pattern on tergites II-V, distinctly fenestrated especially on tergites IV-VI. Lateral taeniae 3 on segments V and VI; 3 to 4 on segment VII and 2 to 4 on segment VIII. Anal spur on tergite VIII with 1 large point and 1 to 3 small apical points. Anal taeniae 54 (44-67).

Remarks

The holotype male was reared from a larva collected from under the rolled edge of *Potamogeton natans* floating leaf. Paratypes are designated as indicated in material examined. The immatures of this species have been identified as *P. ophioides* based on Berg's (1950) work and the adults have often been identified as *P. braseniae*. The species is named after Dr. C.O. Berg.

Biology

Berg (1950) described the biology of this species living on the surface of *Potamogeton natans* floating leaves as follows: they "produce no distinct channels but feed by gnawing away the upper epidermis and mesophyll tissue in irregular patches," and "were found only in the rolled lateral edges of young leaves." We occasionally collected larvae of these species in the rolled edges of older leaves as well as in young leaves, but all reared specimens are from floating leaves of *Potamogeton natans*.

Material Examined

—Reared Specimens

Indiana, Crooked Lake, Noble Co., 20 June 1965, 1 female, 11 July 1965, 3 males, (JES). Sylvan Lake, 29 June 1965, 1 male, (JES). **Minnesota**, marshy pond, Shoreview, Ramsey Co., 45°05'N 93°08'W, June, July, August, 1969-1971, D.E. Maschwitz, 44 males, 11 females. The following specimens from this locality are designates as paratypes: 21 Aug. 1969, 2 males; 17 June 1970, 1 male; 8 July 1970, 3 males, 1 female; 22 July 1970, 4 males; 8 July 1971, 3 males, 2 females; 10 Aug. 1971, 1 male, (UMn). Records from June through August.

—Adult Specimens

Indiana, Big Cedar Lake, N. Columbia City, 21 June 1965, 1 male paratype, (JES). **Massachusetts**, Lexington, 20 July 1961, W.W. Wirth, light trap, 1 male paratype, 1 male, (USNM). **Michigan**, Douglas Lake, Cheboygan Co., 16-18 Aug. 1941, C.O. Berg, slide of hypopygium, designated as an atypical paratype of *P. ophioides* by Townes, (Corn). **Minnesota**, Lower LaSalle Lake, Hubbard Co., 24 July 1970, 1 male paratype, July, August, September 1970, 15 males, New Jersey mosquito light trap, leg E.F. Cook, (UMn). **Missouri**, Wheatland Public Use Area, Elderado Springs, Cedar Co., 6 June 1968, light trap, R.A. Hellenthal, 1 male paratype, (UMn). **Montana**, Rocky Mountain lab, Skalkaho Divide, 22 July 1967, C.B. Philip, malaise trap, 1 male paratype, (JES). **New York**, 1 mi. E Silver Creek, Lake Erie, Chautauqua Co., 19 Sept. 1967, D.W. Webb, 1 male paratype, (INH); Milford Center, 3 Sept. 1935, H.K. Townes, 1 male, (UKan). **Ontario**, Fifth Depot Lake, Wagnersville, 13 July 1967, UV light trap, H.J. Teskey, 1 male paratype, (CNC). Records from June through September.

Polypedilum (P.) braseniae (Leathers) 1922

Chironomus braseniae Leathers, 1922a: 8; Pettit, 1900: 110 (description of unnamed *Cheironomus* (sic) sp. mining in water lilies, probably *P. braseniae*); Leathers, 1922b: 23 (biology, description of all life stages); Johannsen, 1937: 32.
Polypedilum braseniae (Leathers). Townes, 1945: 58; McGaha, 1952: 369; Beck and Beck, 1959: 93; Beck, 1977 (biology); Pinder and Reiss, 1983: 413; (figures of larval characters); Boesel, 1985; Pinder and Reiss, 1986: 429 (figures of pupal characters).

Adult Male

Diagnosis—Very similar to *P. bergi*. Largest species in *illinoense* group; total L over 4.0 mm, often over 5 mm. Phallapodeme relatively straight and narrow, or only slightly expanded at medial end. HR less than 3.0. AR greater than LR. Brachiolum usually with 2 or more setae on at least 1 wing.

Description—(n=8)

Coloration: Head, thorax, abdomen, haltere, and legs uniformly stramineous to light brown. Wing veins stramineous. Pedicel tinged with light brown in some specimens. Flagellum light brown.

Biometrics:	L total 4.80 (4.12-5.20) mm	L thorax 1.31 (1.14-1.51) mm
	L wing 2.48 (2.36-2.76) (5) mm	W wing 0.74 (0.70-0.81) (5) mm

Antenna: AR 1.84 (1.66-2.00) (7). D scape 205 (200-210) (2). L pedicel 125 (120-130) (2). L flagellomeres 1-12 476 (440-500) (7). L flagellomere 13 870 (810-980) (7).

Head: W 660 (630-700). Interocular distance 109 (70-114). HIR 6.06 (4.50-9.43). Vertical S 20.4 (18-24) (5). Clypeal S 24.8 (20-28) (5). Ocelli absent. L maxillary palp segments 1-5: 47 (41-50) (5), 45 (36-52) (5), 122 (113-131) (5), 131 (113-140) (5), 187 (155-207) (5).

Thorax: Antepronotum as in *P. illinoense*. Thoracic sensilla similar to *P. illinoense* but more distinct, postpronotal depression with visible campaniform sensilla in most specimens. Humeral pit very small or absent. A single scutal fossal S on 1 measured specimen. Dorsocentral S 15.8 (12-19) (5), uniserial. Prealar S 5.4 (4-7) (5). Scutellar S in 2 regular rows, anterior row smaller, anterior row 8.3 (5-12) (4), posterior row 12.0 (11-13) (5).

Wing: VR 1.18 (1.17-1.20) (5), wing unspotted. Apices of R_{4+5} and M usually approximately equidistant from the wing apex. R_1 and R_{2+3} usually separated by slight gap apically. Brachiolar S 2.2 (2-3) (5). R_1 S 56.6 (51-63) (5). R_{4+5} S 32.2 (29-34) (5). Squamal S 15.1 (12-19) (7).

Legs: L foretibial apical scale 43 (41-47) (5). L spine on foretibial apical scale 14 (11-16), light, occasionally slightly darkened in some specimens. L mesotibial apical spine 24 (20-27). L hindtibial apical spine 29 (27-32).

	Fe	**T1**	**Ta1**	**Ta2**	**Ta3**
PI	1093 (1050-1170) (6)	797 (740-870) (7)	1253 (1190-1360) (7)	864 (830-910) (7)	671 (640-730) (7)
PII	1248 (1080-1380)	1109 (970-1200)	594 (530-640)	354 (330-370)	265 (250-270)
PIII	1336 (1170-1490)	1298 (1160-1410)	929 (850-990) (7)	539 (510-550) (7)	456 (410-480) (7)

	Ta4	Ta5	LR	BV	SV
PI	524 (490-580) (7)	201 (190-210) (7)	1.59 (1.52-1.64) (7)	1.39 (1.36-1.43) (6)	1.49 (1.42-1.55)
PII	159 (150-170)	99 (90-100)	0.53 (0.52-0.55	3.37 (3.08-3.68)	4.00 (3.89-4.10)
PIII	261 (240-280) (7)	124 (120-130) (7)	0.72 (0.71-0.74) (7)	2.57 (2.46-2.68) (7)	2.82 (2.76-2.89) (7)

Hypopygium: (Figure 56). Tergite IX S 11.4 (4-21), located in broad, middorsal triangular area. Anal point usually curved slightly ventrally; distinctly shorter than IVo, -46 (-32 to -53). SVo similar to *P. falciforme* and *P. bergi*; evenly curved medially; apex bluntly pointed; usually heavier and stouter than in *P. falciforme*. Microtrichia on base numerous and moderately well developed. Basal S 3.4 (3-4). LSVo 102 (90-119). Mean SVoR 1.13 (0.68-1.68). Distal S weak, not exceeding IVo in length. Max L distal S 87 (72-101). IVo narrower relative to its length compared to other species in *illinoense* group; medial margin usually straight, subparallel to lateral margin. L IVo 87 (72-101). W IVo 44 (29-65) (7). Dorsal S 14.6 (11-19), not concentrated on lateral margin. Apex usually with 1 or 2 prominent setal bearing papillae. L gonostylus 201 (185-214), generally wider than in other *illinoense* group species. GsR 4.10 (3.48-4.75). Medial S 4.6 (4-6), well developed. Max L medial S 75 (63-86). Max L apical S 25 (23-29) (7). HR 2.54 (2.23-2.94). L posterior lobe gonocoxite 81 (70-90). HV 2.40 (2.24-2.56). Phallapodeme straight, not expanded, or only slightly expanded, at medial end. Medial end, and sometimes lateral end also, rounded or knob-like.

Type locality: Holotype male, **New York**, Spencer Lake, near North Spencer, July 1915, reared from leaf of *Brasenia peltata* (Corn).

Larva

Diagnosis—Mentum with 8 to 14 apparent teeth, central portion deeply concave in specimens from *Nymphaea odorata*; truncate with 2 small central teeth in specimens from *Braseniae schreberi*.

Description—(n=3; 2 males, 1 female)

Live color: Light greenish orange.

56

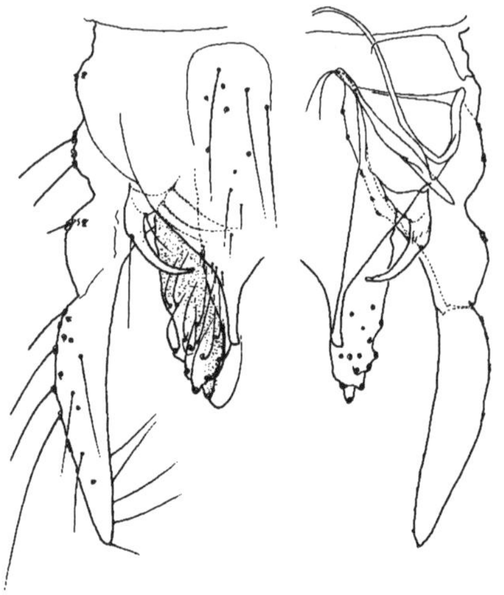

57

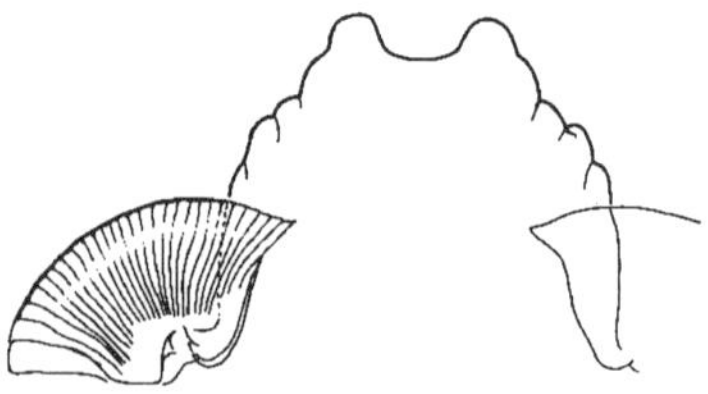

58

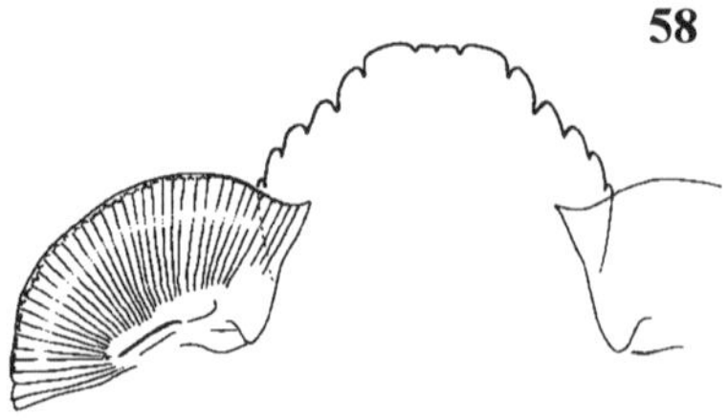

Figures 56-58. *P. braseniae* (Leathers): 56- hypopygium; 57-mentum of specimen from *Nymphaea odorata*; 58-mentum of specimen from *Brasenia schreberi*.

Eye spots: Nearly contiguous (Leathers, 1922b).

Biometrics: L total 6.30 (6.00-6.60) mm (2) L head capsule 318 (304-326)

Mentum: (Figures 57, 58). W 158 (155-162). Menta of specimens from *N. odorata* usually show good deal of wear and many have no distinct teeth; central portion deeply concave. Mentum with 8 to 10 apparent teeth. Teeth rounded and approximately equal in size except for very large first lateral teeth. Specimens from *B. schreberi* with 14 teeth (other differences in specimens from *B. schreberi* discussed below).

Ventromental plates: W 116 (113-122). L 68 (63-72), without posterior lobe. VmPR 1.73 (1.66-1.79). Striations 22 (20-23), very distinct except for narrow band along curved anterior margin. Distance between ventromental plates 105 (104-106). VmPSR 1.11 (1.09-1.13).

Antenna: (Figure 59). L segments 1-5: 69 (68-70) (4), 22 (22-23) (4), 15 (14-18) (4), 12 (11-14)(4), 5 (5-5). AR 0.81 (0.77-0.87). L primary blade on segment 1 54 (50-59) (2). L secondary blade 11 (10-13) (2). Lauterborn organs small, difficult to see in some specimens, not reaching beyond basal one-fourth of antennal segment 3; arch shaped. L Lauterborn organs 6 (6-6) (2), triangular structures visible at base. L PS 11 (10-13) (2). ROR 4.64 (4.29-5.07). Two sense pits present on segment 1.

Labrum-epipharynx: Labial setae I and II slightly expanded laterally; lateral margins roughened or with serrations. PE in three distinct parts, each with 1 large tooth (Figure 61). LL difficult to see but apparently without teeth. Premandibles broad and curved, simple as in Figure 60 or bifid.

Mandible: L 219 (209-234), with three teeth. Teeth very blunt and rounded, possibly due to wear. Medial margin apparently without serrations. L seta subdentalis 23 (1), rod-like with blunt tip. PMa not seen.

Maxilla: Lacinia distinctly bifid, anterior branch heavier than posterior branch. L lacinia 65 (63-65). Galea very prominent, about two-thirds as long as maxillary palp. L maxillary palp segment 1 42 (41-43). Segments 2 and 3 small.

Larvae from *B. schreberi*: Larvae feeding on *B. schreberi* differ from the above description as follows (n=2). Mentum smaller, W 140 (135-144), with 14 teeth. Median 2 teeth small, first laterals very large and broad. Each of succeeding lateral teeth 2-5 progressively smaller, last lateral teeth very small. Median 4 teeth blunt, of even height, apparently due to wear (Figure 58). Ventromental plates, wider and generally larger, W 139 (135-142), L 76 (68-83). Distance between ventromental plates smaller, 82 (77-88), VmPSR 1.68 (1.59-1.76). Antenna similar except segment 1 shorter, L 62 (59-65) (3). ROR 3.49 (3.11-3.87). Galea of maxilla not as large.

59

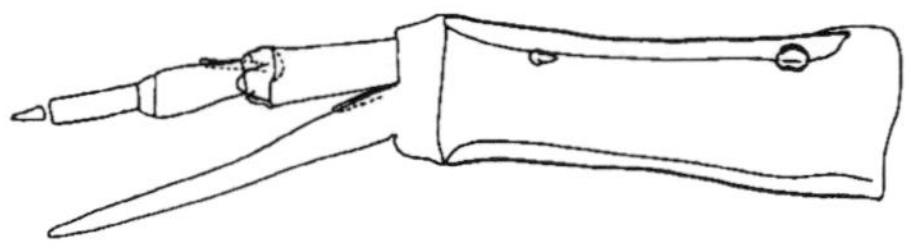

60

61

Figures 59-61. *P. braseniae* (Leathers): 59-antenna; 60-premandible; 61-pecten epipharyngis.

Pupa

Diagnosis—Anal spur on segment VIII absent. Lateral taeniae on segments V and VI very small compared to other species in *P.* (*Polypedilum*) and *P.* (*Uresipedilum*).

Description—(n=5, all males)

Biometrics: L total 5.53 (5.22-5.81) (4) mm L cephalothorax 1.44 (1.40-1.49) (4) mm

Coloration: Typical.

Abdomen: (Figure 62). Pedes spurii A absent. Pedes spurii B moderately developed on segment I, poorly developed or absent on segment II. Hooks slightly larger than large anterior spinules on tergite III. Intersegmental spinules hook-like in appearance, pointing anteriorly, uniserial or partially biserial, slightly larger than large anterior spinules on tergites IV and V. Shagreen spinules covering most of tergites II-V, forming a rough triangle on tergite II. Fenestrations absent (see Remarks). Anterior spinules on tergite VII arranged in rows. Lateral taeniae on segments V and VI, and anterior 2 on segment VII, very small. Posterior 2 on segment VII and those on segment VIII moderately small. Anal spurs absent. Anal taeniae 35.8 (33-38) on each anal lobe.

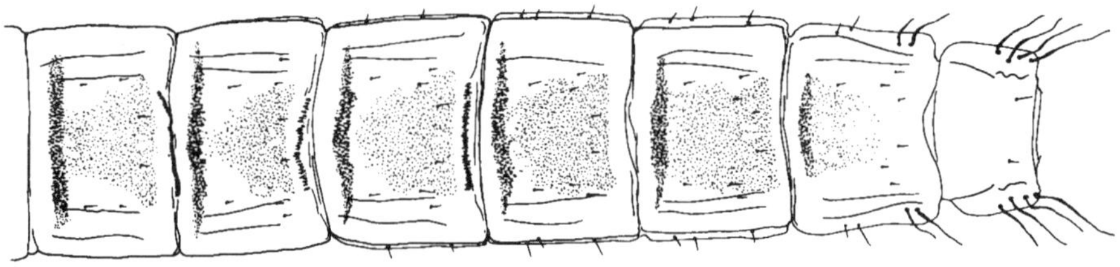

Figure 62. *P. braseniae* (Leathers), pupal abdomen.

Remarks

Larvae and pupae of this species have several unique characters not found on larvae and pupae of other species in the genus (see comments of Pinder and Reiss, 1983: 335 and 1986: 346). However, the adults clearly fit the *illinoense* group and are very similar to *P. bergi* and *P. falciforme*. *Polypedilum braseniae* highlights the problem of divergence of specific, and even generic, associations based on the characters of the various life stages. To further confuse the issue, distinct differences exist in the immature stages, especially the larvae, between specimens taken from *Nymphaea odorata* versus *Brasenia schreberi*. Adults reared from larvae from the two plants appear to be the same. It is not known whether the larval differences are an environmental artifact, possibly due to differences the two plants have on the extent of wear to the mentum teeth, or whether genetic differences are involved. Pupae from *B. schreberi* are the same as described above except for distinct fenestrations in the shagreen spinules on tergites IV-VII. All the *P. braseniae* specimens we examined appeared to have simple premandibles; however, Epler (1995) reports that most of the *P. braseniae* larvae he examined had bifid premandibles.

Biology

The larvae feed on and channel the top surface of the floating leaves of a variety of aquatic plants, but seem to prefer *Brasenia schreberi* and *Nymphaea odorata*. Specimens were reared from both plant species. Leathers (1922b) discussed the biology of this insect in detail, and Beck (1977) summarized its environmental requirements.

Material Examined

—Reared Specimens

Minnesota, marshy pond, Shoreview, Ramsey Co., 45°05'N 93°08'W, 25 May 1970, 2 males, 1 female; 9 June 1971, 3 males; several dates in July 1968, 1969, 1970, 9 males, 3 females, plus many larvae and pupae unassociated with adults, D.E. Maschwitz, (UMn); Walker Brook Lake, 5 mi. SE Bagley, Clearwater Co., 47°29'N 95°18'W, 16 June 1969, 1 male, 1 female, plus 8 larvae, 8 pupae not associated with adults, D.E. Maschwitz, (UMn); Squaw Lake, Itasca State Park, Clearwater Co. , 47°16'N 95°16'W, 16 Aug. 1971, 2 males, 2 females, D.E. Maschwitz, (UMn); Snail Lake, Shoreview, Ramsey Co., 45°04'N 93°08'W, 9 June 1971, 2 males, D.E. Maschwitz, (UMn). **Ontario**, Sasjawarn Lake, Algonquin Park, 6 June 1967, 1 male, D.M. Wood and J. Martin, (FRB).

—Additional immature specimens not associated with adults

Minnesota, Deep Lake, North Oaks, Ramsey Co., 45°07'N 93°05'W, 19 June 1969, 1 larva and pupa, D.E. Maschwitz, (UMn). **New York**, Ithaca, 3 pupae, (Corn). **Manitoba**, Lily Pad Lake, Whiteshell Provincial Park, S half sec. 30, R 17E T 10N, 14 June 1968, 5 larvae, D.E. Maschwitz (UMn). **Ontario** 35 mi. S Wawa, 17 June 1967, J. Martin, 1 larva, 2 pupae, (JES). Records rom May through August.

—Adult Specimens

Holotype male, (slide of hypopygium). **Michigan**, Cheboygan Co., 1 Aug. 1947, 3 males, ex *Nymphaea odorata*; 7 Aug. 1947, 1 male, ex *Brasenia schreberi*, Y.J. McGaha, (USNM). **Minnesota**, Blue Mounds State Park, Rock Co., 28 May 1970, New Jersey mosquito light trap, 1 male, leg. E.F. Cook, (UMn). **South Dakota**, Platte Bay, Lake Francis Case, Platte, 25 May-1 June 1967, P. Hudson, 1 male, (JES). Records from May, June, and August.

Polypedilum (*P.*) *calopterum* (Mitchell) 1908

Chironomus calopterus Mitchell, 1908: 8 (female); Johannsen, 1908: 276 (key)
Chironomus (*Stenochironomus*) *calopterus* Johannsen, 1937: 28.
Polypedilum fallax var. *calopterus* (Mitchell), Townes, 1945: 54.
Polypedilum calopterum (Mitchell), Boesel, 1985: 257 (recognition of species distinct from *P. fallax*, figure of hypopygium); Oliver *et al.*, 1990: 51 (listing of species); Oliver and Dillon, 1994: 10; Caldwell *et al.*, 1997.

Remarks

Polypedilum calopterum was originally described from a female and was distinguished by the presence of a broad, dark band through the middle of the wing. Townes (1945) could not distinguish two males of *P. calopterum* from *P. fallax* and designated the former a variety of the latter. Boesel (1985) collected both sexes of *P. fallax* and *P. calopterum* in the same environments and concluded that the two forms are distinct. He recognized *P. calopterum* (Mitchell) as a separate species. The two species are separated in Boesel's (1985) key to adults by the presence of a double row of dorsolateral (dorsocentral) setae on the scutum of *P. calopterum* and a single row on *P. fallax* (and *P. artifer*). *Polypedilum calopterum* is a larger species, wing length 3.8 mm; *P. fallax* wing length, 2.9 mm. The hypopygium of *P. calopterum* (Boesel, 1985: 248) appears to be indistinguishable from the hypopygia of *P. fallax*, *P. artifer*, and *P. walleyi*. (Boesel (1985) synonymized *P. walleyi* with *P. calopterum*). Caldwell *et al.* (1997) did not recognize *P. calopterum* as a synonym of *P. fallax*. We could find no differences between the immatures of *P. fallax* and *P. calopterum*. *Polypedilum calopterum* is not included in the keys.

Material Examined—Reared Specimens

Quebec, pool in Lusk Cave, Gatineau Park, Jon Martin & A.L. Hamilton 22 July 1966, 4 females, 29 July 1966, 2 females, 12 July 1966, 1 male, (FRB and CNC).

Polypedilum (*P.*) *castornama* New Species

Adult Male

Diagnosis—Very similar to *P. laetum* except for following hypopygial differences. SVo base with large tubercle curving medially and bearing 2-4 setae on apex. Distal seta on SVo weak. Gonostylus larger and medial setae usually more numerous than in *P. laetum*.

Description—(n=5)

Coloration: As in *P. laetum* except forefemur lighter proximally, gradually becoming darker brown distally. Other femora more uniformly brown.

Biometrics:	L total 4.49 (4.29-4.68) mm	L thorax 1.18 (1.12-1.24) mm
	L wing 2.47 (2.39-2.53) mm	W wing 0.72 (0.66-0.77) mm

Antenna: AR 1.66 (1.55-1.86). D scape 178 (170-180) (4). L pedicel 140 (140-140) (2). L flagellomeres 1-12 492 (440-530). L flagellomere 13 812 (800-820).

Head: W 597 (580-610) (3). Interocular distance 107 (100-110) (3). HIR 5.61 (5.27-6.10) (3). Vertical S 15.3 (13-20). Clypeal S 20 (17-26). Ocelli very small. L maxillary palp segments 1-5: 50 (47-56), 39 (36-43), 165 (158-171) (4), 151 (140-155), 217 (214-223) (4).

Thorax: Dorsal edge of antepronotum as in *P. laetum* in lateral view. Postpronotal depression as in *P. laetum*. Antepronotum without setae. Dorsocentral S 16.2 (13-19). Scutal fossal S absent, but anterior most dorsocentral seta placed well forward in 2 specimens. Dorsocentral setae uniserial anteriorly, biserial medially and posteriorly. Prealar S 5.7 (5-7). Scutellar S 17.6 (16-20), usually in 2 rows with anterior row smaller.

Wing: VR 1.07 (1.04-1.10). With 3 usually distinct wing spots. Placement of spots as in *P. laetum* except spots at apices of M_{3+4} and Cu absent. Apices of R_{4+5} and M about equidistant from wing apex. R_1 and R_{2+3} separated slightly apically. Brachiolar S 1. R1 S 41.2 (38-50). R_{4+5} S 35.2 (30-45), Squamal S 16.8 (13-21).

Legs: L foretibial apical scale 41 (36-47), scale with apical spine, apical 2/3 of spine darkened. L spine 14 (7-20). L mesotibial apical spine 19 (16-20). L hindtibial apical spine 23 (18-30) (4).

	Fe	T1	Ta1	Ta2	Ta3
PI	1020 (1010-1040)	840 (820-860)	1136 (1110-1170)	680 (650-690)	526 (500-540)
PII	1140 (1090-1170)	956 (940-970)	592 (560-610)	336 (330-340)	256 (250-260)
PIII	1234 (1210-1250)	1088 (1060-1120)	872 (830-910)	476 (460-490)	376 (360-390)

	Ta4	Ta5	LR	BV	SV
PI	380 (340-400)	168 (160-170)	1.37 (1.34-1.39)	1.72 (1.65-1.81)	1.63 (1.61-1.66)
PII	168 (160-180)	106 (100-110)	0.62 (0.60-0.63)	3.12 (3.06-3.21)	3.56 (3.49-3.64)
PIII	230 (220-250)	128 (120-130)	0.80 (0.78-0.81)	2.66 (2.62-2.69)	2.69 (2.63-2.78)

Hypopygium: (Figures 63, 64). Tergite IX S 10.4 (9-13). Apex of tergite IX at base of anal point truncate, or, in some specimens, concave on either side of anal point. Anal point tapered slightly from base to apex; shorter than IVo, -18 (-2 to -27). Base of SVo with posteromedial tubercle. Tubercle occasionally ventral to and hidden by SVo (Figure 64). Proximal portion of base always with one seta. Apex of tubercle with 2.6 (2-4) setae. Distal portion of SVo similar to *P. laetum*. Distal seta small, often not exceeding apex of anal point; max L distal S 75 (68-88). LSVo 108 (99-122). Mean SVoR 0.89 (0.63-1.04). IVo as in *P. laetum*. L IVo 152 (144-164). W IVo 31 (26-36). IVo dorsal S 12.2 (7-16). Gonostylus generally larger than in *P. laetum*, L 153 (149-162). Gonostylus medial margin usually straight, lateral margin broadly rounded. Medial S 7.0 (5-8), generally more numerous than in *P. laetum*; secondary row of medial S present, nearly as large as those in main row. Max L medial S 58 (56-61). Apical S large, L 57 (52-59) (4). L gonostylus 153 (149-162). GsR 2.95 (2.65-3.43). L gonocoxite posterior lobe 71 (59-83). HR 2.20 (1.81-2.65). HV 2.95 (2.78-3.14). Microtrichia on basal plate; phallapodeme similar to *P. laetum*.

Type locality: Holotype male, **Minnesota**, Beaver Creek Valley State Park, Houston Co., 6 October 1970, leg. E.F. Cook, New Jersey mosquito light trap, (UMn).

63

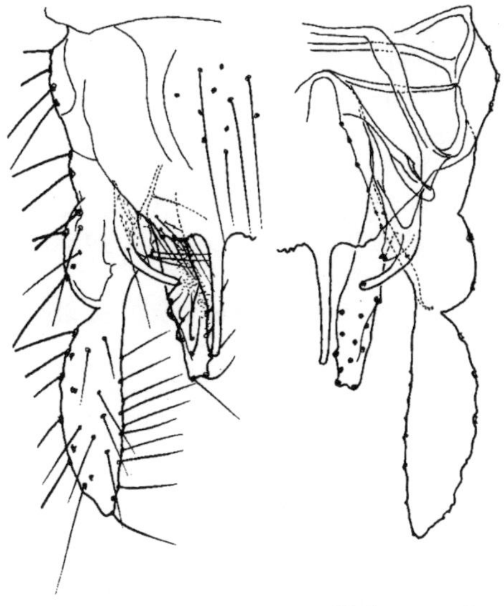

64

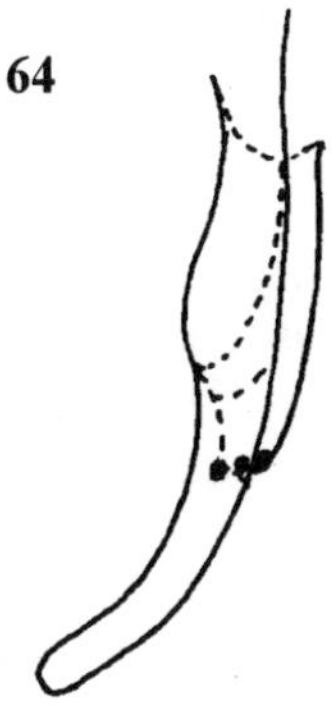

Figures 63-64. *P. castornama* sp. nov.: 63-hypopygium; 64-variation of superior volsella.

Remarks

All the remaining specimens examined are designated as paratypes. At present this species is known only from the type locality, for which the species is named. The immature stages are unknown. Rossaro (1985:11) discusses the variability in number and placement of setae on the SVo in European *P. laetum* and feels that the SVo of *P. castornama* falls within that variability. He suggests that *P. castornama* and *P. laetum* are the same species.

Material Examined—Adult Specimens

Minnesota, Beaver Creek Valley State Park, Houston Co., 16 June, 1970 one male; 30 June 1970, 10 males; 8 July 1970, 1 male; 14 July 1970, 26 males; 23 July 1970, 1 male; 28 July 1970, 3 males, New Jersey mosquito light trap, leg. E.F. Cook (UMn).

Polypedilum (U.) cinctum Townes 1945

Polypedilum cinctum Townes, 1945: 59; Oyewo and Sæther, 1998.

Adult Male

Diagnosis—IVo elongated posteriorly, extending beyond apex of anal point. Anal point not expanded laterally except slightly near apex. SVo boot-shaped as in *P. obtusum*. Central portion of abdominal tergites I-VI brown.

Description—(n=7)

Coloration: Head, pedicel, and flagellum light brown. Thoracic vittae, preepisternum II, anepisternum II, epimeron II, and postnotum light brown to brown, other parts of thorax stramineous to light brown in darker specimens. Haltere and legs stramineous. Central parts of abdominal tergites I-VI or VII light brown. Hypopygium stramineous.

Biometrics: L total 4.02 (3.94-4.34) (6) mm; L thorax 1.15 (1.09-1.22) (6) mm; L wing 2.19 (2.10-2.29) mm; W wing 0.68 (0.67-0.68) (5) mm

Antenna: AR 1.86 (1.67-2.00) (6). D scape 155 (150-160) (6). L pedicel 115 (110-120) (2). L flagellomeres 1-12 391 (380-400). L flagellomere 13 733 (670-780) (6).

Head: W head 538 (500-560) (5). Interocular distance 112 (100-120) (5). HIR 4.81 (4.58-5.00) (5). Vertical S 14.1 (12-16) (4). Clypeal S 18.8 (14-22) (6). Ocelli present. L maxillary palp segments 1-5: 36 (32-41) (4), 34 (32-38) (4), 158 (151-162) (3), 155 (142-164) (3), 238 (227-257) (3).

Thorax: Prothorax as in *P. illinoense*. Postpronotal depression small, campaniform sensilla weakly visible. Humeral pit present. Dorsocentral S 15.7 (12-19) (5), uniserial except at posterior end in some specimens. Scutal fossal S absent. Prealar S 5.4 (4-6) (5). Scutellar S in 2 usually regular rows. Setae in 2 rows nearly same size; anterior row S 8.2 (4-12) (5), posterior row S 14.8 (11-17) (5).

Wing: VR 1.15 (1.13-1.18). Apices of R_{4+5} and M approximately equidistant from wing apex. R_1 and R_{2+3} adjacent apically. Brachiolar S 1. R_1 S 42.2 (37-47). R_{4+5} S 42.7 (36-55). Squamal S 18.0 (16-20) (5).

Legs: Foretibial apical spine absent. L foretibial apical scale 32 (25-36). L mesotibial apical spine 18 (16-20) (6). L hindtibial apical spine 18 (16-18) (6).

	Fe	T1	Ta1	Ta2	Ta3
PI	867 (830-940)	687 (650-740)	978 (920-1020) (6)	670 (620-710) (6)	510 (450-550) (6)
PII	1027 (980-1080)	842 (790-880)	523 (500-550) (6)	280 (260-290) (6)	208 (190-220) (6)
PIII	1057 (1020-1100)	968 (920-1030)	746 (720-780) (6)	371 (350-390) (6)	328 (310-340) (6)

	Ta4	Ta5	LR	BV	SV
PI	383 (330-420) (6)	170 (160-180) (5)	1.41 (1.38-1.50) (6)	1.46 (1.43-1.55) (5)	1.59 (1.53-1.63) (6)
PII	133 (120-150) (6)	85 (80-90) (6)	0.61 (0.58-0.63) (6)	3.43 (3.24-3.56) (6)	3.62 (3.47-3.76) (6)
PIII	198 (190-210) (6)	98 (90-100) (6)	0.76 (0.74-0.78) (6)	2.80 (2.66-2.91) (6)	2.74 (2.66-2.80) (6)

Hypopygium: (Figure 65). Tergite IX S 16.0 (12-19) (6), located in broad middorsal egg-shaped area. Anal point curved ventrally, not expanded laterally except slightly at apex. Anal point not exceeding posterior end of IVo, -25 (-14 to -34). SVo boot-shaped as in *P. obtusum*, apical setate tubercle reduced in some specimens. Microtrichia present at base of distal seta. Basal S 2.0 (2-2). LSVo 100 (92-108). Mean SVoR 0.98 (0.79-1.15). Max L distal S 86 (65-108). IVo long, distal portion narrowed, apex truncate. Medial margin straight, lateral margin irregular with mid portion expanded laterally. Apex with posteriorly directed seta. Dorsal surface free of microtrichia except apically and along medial margin. L IVo 147 (140-158). W IVo 40 (36-47). IVo dorsal S 22.2 (16-29), moderately developed. Gonostylus as in *P. flavum*. Medial S 8.4 (7-10), weakly developed, and often 1 or more setae out of line with the main row. Max L medial S 41 (29-47). Max L apical S 27 (20-34) (5). L gonostylus 184 (160-203). GsR 3.05 (2.83-3.46). HR variable, 3.13 (2.42-4.04). L posterior lobe gonocoxite 60 (50-79). HV 2.21 (1.95-2.57) (6). Phallapodeme straight, only slightly expanded medially.

Type locality: Holotype male, **Nevada**, Reno, 5 Aug. 1915, M.G. Dyar, male, (USNM).

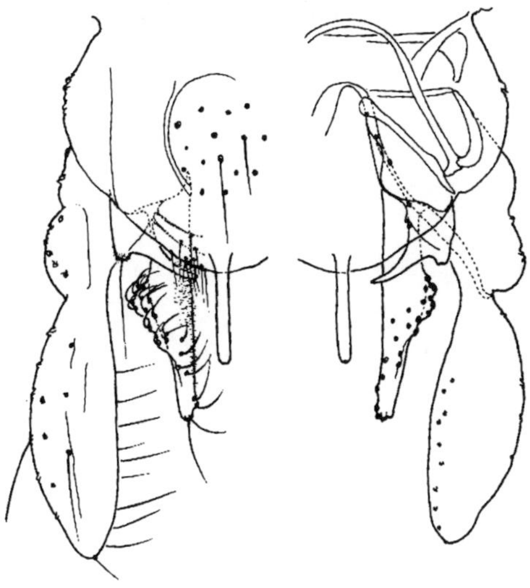

Figure 65. *P. cinctum*, hypopygium.

Material Examined—Adult Specimens

Nevada, Holotype male (slide of hypopygium); Reno, 6 July 1916, 5 and 6 Aug. 915, 3 paratype males, M.G. Dyar (USNM). **California**, Castle Lake, Siskiyou Co., 14 July 1953, 3 males, (JES). **Oregon**, Salem, Willamette R., 22 May 1963, K. Goeden, light, 14 male, (USNM). **Virginia**, Elkton, 13 June 1958, E.M. Raffensperger, males (USNM). Records from May through August.

Polypedilum (*P.*) *crassum* New Species

Adult Male

Diagnosis—SVo and gonostyli stout and heavy compared to other species in *laetum* group, SVo distal seta located near base. SVo basal setae borne on small tubercles. Small antepronotal setae present. No spine on apical scale of foretibia. Hypopygium similar to *P. trigonus* but distinguished from this species by larger gonostyli and lack of large microtrichia on SVo base.

Description—(n=2)

Coloration: Pedicel, head, thorax, abdominal tergites brown to dark brown. Legs light brown, proximal 1/3 of mid and hind femora darker. Haltere, flagellomeres, palps, abdominal sternites brown to light brown. Wing veins stramineous.

Biometrics:	L total 4.64 (4.51-4.77) mm	L thorax 1.23 (1.22-1.23) mm
	L wing 2.26 (2.13-2.38) mm	W wing 0.62 (0.57-0.67) mm

Antenna: AR 1.43 (1.41-1.44). D scape 170 (160-180). L pedicel 120 (1). L flagellomeres 1-12 480 (460-500). L flagellomere 13 685 (650-720).

Head: W 600 (580-620). Interocular distance 100 (80-120). HIR 6.21 (5.17-7.25). Clypeal S 28 (25-30). Vertical S 13 (12-14). Ocelli absent. L maxillary palp segments 1-5: 44 (41-47), 39 (36-43), 132 (122-142), 154 (151-158), 236 (223-250).

Thorax: Dorsal edge of antepronotum developed anteriorly but not projecting beyond anterior margin of scutum. Antepronotum sensillum very small; other antepronotal sensilla similar to *P. laetum*. Antepronotal S 1.5 (1-2). Dorsocentral S 16 (16-16), uniserial, becoming biserial posteriorly. Scutal fossal S absent. Prealar S 6.5 (6-7). Scutellar S 13.5 (13-14), in single row.

Wing: VR 1.19 (1.17-1.20). Wing spots absent. Apex of R_{4+5} slightly to distinctly closer to wing apex than apex of M. R_1 and R_{2+3} slightly separated apically. Brachiolar S 1. R_1 S 53 (50-56). R_{4+5} S 39 (37-40). Squamal S 13 (13-13).

Legs: Foretibial apical scale without a spine, L scale 51 (50-52) (Figure 18). L mesotibial spine 23 (23-23). L hindtibial spine 26 (25-27).

	Fe	Tl	Ta1	Ta2	Ta3
PI	935 (890-980)	725 (670-780)	—	—	—
PII	1070 (1020-1120)	950 (900-1000)	540 (510-570)	310 (290-330)	230 (220-240)
PIII	1120 (1060-1180)	1125 (1070-1180)	810 (1)	470 (1)	340 (1)

	Ta4	Ta5	LR	BV	SV
PI	—	—	—	—	—
PII	145 (140-150)	105 (100-110)	0.57 (0.57-0.57)	3.25 (3.25-3.25)	3.76 (3.74-3.78)
PIII	190 (1)	130 (1)	0.68 (1)	2.82 (1)	2.94 (1)

Hypopygium: (Figure 66). Tergite IX S 13 (1), located in somewhat egg-shaped lighter area. Apex of tergite IX rounded at base of anal point. Anal point not tapered except near base, apex about even with apices of IVo, -1 (-18 to 16). SVo large and stout, angled medially at point of distal seta origin, apex with rounded knob. Distal S moderately well developed, placed near the base. A few small microtrichia on SVo around origin of distal seta. Max L distal S 103 (99-106). LSVo 94 (88-99). Mean SVoR 2.07 (1.75-2.58). Basal setae located in a group and borne on small tubercles, basal S 2.8 (2-3). IVo long, angled medially with posteriorly directed apical seta. L IVo 145 (135-155). W IVo 29 (27-29). IVo dorsal S 15.5 (15-16). Gonostylus large compared to other *laetum* group species, lateral margin broadly rounded. L gonostylus 156 (155-158). GsR 2.48 (2.46-2.50). Medial S 6.8 (6-8), strong. Max L medial S 75 (72-77). L apical S 52 (47-56). Posterior lobe of gonocoxite very large; basal plate distinct. L posterior lobe gonocoxite 78 (70-86). HR 2.04 (1.82-2.26). HV 2.97 (2.84-3.10). Phallapodeme slightly expanded toward medial end.

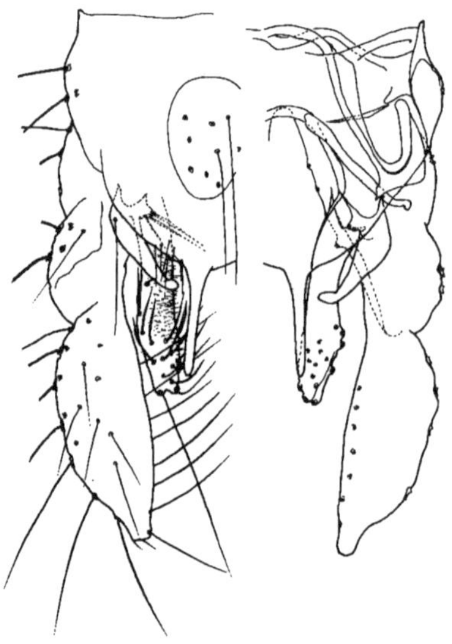

Figure 66. *P. crassum* sp. nov., hypopygium.

Type locality: Holotype male, **Minnesota**, Univ. of Minnesota Forestry and Biological Station, Itasca State Park, 18 June 1968, D.E. Maschwitz, at light, (UMn). Paratype male, **Minnesota**, New London Fish Hatchery, Kandiyohi Co., 2 June 1969, New Jersey mosquito light trap, leg. E.F. Cook, (UMn).

Remarks

This species is known only from these two specimens, and is named for the large stout superior volsella and gonostylus.

Polypedilum (U.) cultellatum Goetghebuer 1931a

Polypedilum cultellatum Goetghebuer, 1931a: 212; Kruseman, 1933: 134 (description of *P. convictum* but figure 11 of hypopygium looks like *P. cultellatum* (?); Goetghebuer, 1936: 74 (biology); Goetghebuer, 1937: 60, Brundin, 1947: 64 (hypopygium figured); Brundin, 1949: 769 (distribution and biology); Thienemann, 1954: several pages (biology); Dittmar, 1955; 472 (biology); Lehmann, 1971: 500 (biology); Pinder, 1974: 197 (figure of hypopygium); Langton, 1984: 245 (figure of pupa); Rossaro, 1985: 21 (figure of hypopygium, larval and pupal characters); Ashe and Cranston, 1991 (Palaearctic species); Niitsuma, 1992 (description, figure of hypopygium); Sasa and Kikuchi, 1995 (figure of hypopygium); Oyewo and Sæther , 1998.

Polypedilum subcultellatum Sublette, 1960: 204 new synonymy.

Adult Male

Diagnosis—SVo broad, rounded apex bearing 3 to 5 setae. SVo with preapical digitiform process on medial side similar to *P. flavum*. Anal point relatively short, narrow and tapered apically.

Description—(n=4)

Coloration: (Taken mainly from single slide-mounted specimen of *P. subcultellatum*). Thoracic vittae and postnotum brown. Preepisternum, scutellum, head, flagellum, and legs light brown, pedicel shade lighter. Abdominal tergites light brown, posterior third of tergites I through VI shade darker (not visible in uncleared European specimens available to us). Wing veins light brown, haltere stramineous.

Biometrics:	L total 4.16 (3.70-4.59) (3) mm	L thorax 1.12 (1.07-1.19) (3) mm
	L wing 2.31 (1.99-2.62) (2) mm	W wing 0.65 (0.57-0.73) (2) mm

Antenna: AR 1.65 (1.51-1.73) (3). L flagellomeres 1-12 467 (420-490) (3). L flagellomere 13 770 (720-850) (3).

Head: Vertical S 15.3 (15-16) (3). Clypeal S 21 (1). L maxillary palp segments 2-5: 50 (1), 119 (1), 113 (1), 191 (1).

Thorax: Dorsal edge of antepronotum as in *P. illinoense*. Prothoracic campaniform sensillum and humeral pit apparently absent. Postpronotal depression small, without visible campaniform sensilla. Antepronotum without setae. Scutal fossal S 2.3 (1-3) (3). Dorsocentral S 18.0 (16-21) (3), uniserial, occasionally irregular. Prealar S 6.3 (5-7) (3). Scutellar S in 2 rows, anterior row with smaller setae; anterior rows 3.5 (3-4) (2); posterior row S 13.5 (13-14) (2).

Wing: VR 1.16 (1.14-1.17) (2). Wing unspotted. Apices of R_{4+5} and M about equidistant from wing apex. R_1 and R_{2+3} adjacent apically. Brachiolar S 1 (1). R_1 S 53.0 (51-55) (2). R_{4+5} S 50.5 (46-55) (2). Squamal S 15.0 (12-19).

Legs: Foretibial apical scale triangular in shape, narrowing apically to sharp point, without apical spine. L foretibial apical scale 50 (50-50) (2).

	Fe	Tl	Ta1	Ta2	Ta3
PI	1020 (930-1110) (2)	725 (650-800) (2)	104 (1)	73 (1)	—
PII	1160 (1040-1280) (2)	975 (880-1070) (2)	550 (520-580) (2)	365 (330-400) (2)	265 (240-290) (2)
PIII	1165 (1000-1330) (2)	1150 (1040-1260) (2)	920 (1)	520 (1)	440 (1)
	Ta4	**Ta5**	**LR**	**BV**	**SV**
PI	—	—	1.62 (1)	—	—
PII	165 (160-170) (2)	95 (80-110) (2)	0.57 (0.54-0.59) (2)	3.03 (3.02-3.04) (2)	3.90 (3.71-4.09) (2)
PIII	250 (1)	120 (1)	0.72 (1)	2.65 (1)	2.84 (1)

Hypopygium: (Figure 67). Tergite IX S 4.7 (3-7) (3), located in poorly defined, narrow, middorsal area. Anal point short, straight, and tapered apically, usually terminating well short of IVo apices, -23 (-5 to -34). SVo with subapical medial digitiform process. Apex of SVo rounded, bearing 4.4 (3-5) setae. SVo, except for medial process, densely covered with microtrichia. SVo without distinct basal portion. Basal S 2.3 (2-3). LSVo and SVoR not determined. Max L distal S 56 (34-63), usually not exceeding IVo posteriorly. IVo sides subparallel, apical 1/3 to 1/4 moderately expanded dorsally, or, if IVo flattened or rotated slightly, IVo appears slightly expanded laterally, Most dorsal setae inserted on this swelling; swelling free of microtrichia. Apical posteriorly directed seta present. L IVo 121 (115-129). W IVo 30 (29-32) (3). IVo dorsal S 17.1 (11-20), moderately well developed. Gonostylus narrow, tapered to blunt point apically, medial margin generally straight. Medial S 5.6 (5-6), in straight row. Max L medial S 51 (41-59). Max L apical S 26 (25-27) (2). L gonostylus 183 (155-214). GsR 4.31 (3.45-4.88). HR 2.64 (2.36-3.06). L posterior lobe gonocoxite 70 (56-79). HV 2.34 (2.20-2.42) (3).

Figure 67. *P. cultellatum* Goetghebuer, hypopygium (uncleared specimen, internal structures not visible).

Type locality: *P. cultellatum*, **Belgium**, 10 Sept 1931, male, (? Brussels Museum); *P. subcultellatum*, **California**, 6 mi. E. Orosi, 8 July 1947, male, (USNM).

Larva
See figures of larva in Rossaro (1985).

Pupa
Diagnosis—Cephalothorax with prominent tubercle anterior to wing sheath as in *P. falciforme* and *P. bergi*, but spinules relatively much larger and darker than in these two species. Usually distinguished from *P. falciforme* by its larger size, and from *P. bergi* by more extensive spinule pattern on tergite II.

Description (n=3; sex unknown)
Biometrics: L total 5.47 (5.23-5.70) (2) mm L cephalothorax 1.39 (1.33-1.45) (2) mm

Coloration: Typical.

Abdomen: (Figure 68). Pedes spurii B on segments I and II weakly to moderately developed. Pedes spurii A present. Hooks on tergite II smaller than large anterior spinules on tergite III. Large band of anterior spinules present on tergites II through VI. Spinules covering most of tergites II and III, except for large fenestrations and anterolateral corners of tergite III. Spinules on tergites IV through VI forming broad triangular or "V" shaped pattern, with fenestrations. Intersegmental spinules between segments III and IV uniserial to irregularly biserial, similar in shape to hooks when uniserial; spinule-form when double. Intersegmental spinules between IV and V irregularly uniserial to biserial, spinule-form. Both intersegmental spinule rows with spines pointing anteriorly. Spinules on tergites VII and VIII small, usually arranged in rows. Three lateral taeniae on segments V and VI, setae of three sizes, anterior seta smallest and posterior seta largest. Segments VII and VIII with 4 lateral taeniae, of about the same size. Anal spur often without smaller apical accessory points, occasionally with 1 or 2. Anal taeniae 37.4 (26-45) (4) on each anal lobe.

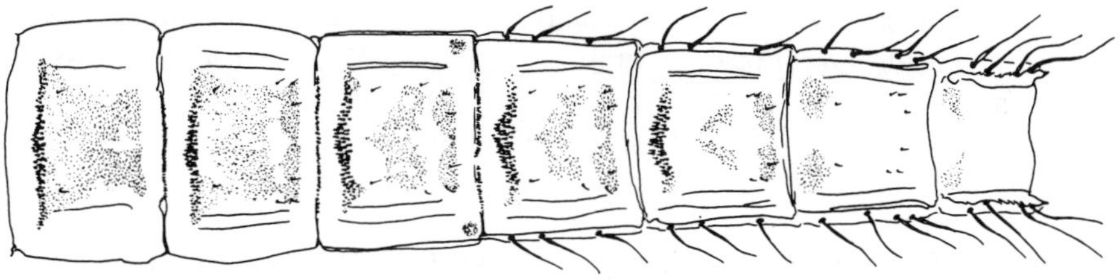

Figure 68. *P. cultellatum* Goetghebuer, pupal abdomen.

Remarks
Rossaro (1985) includes figures of both the larva and pupa, and Langton (1984) figures the pupa.

The presence of several apical setae on the SVo separates *P. cultellatum* from all other Nearctic species of *P.* (*Polypedilum*), but, because our description is based on very limited material, it very likely does not reflect the range of variation within the species. Despite the lack of specimens (*P. subcultellatum* is known from only two specimens), *P. subcultellatum* Sublette is considered a synonym of *P. cultellatum*. In our specimens and in the figure of the hypopygium by Brundin (1947: Figure 97), the inferior volsella is not expanded laterally nearly as much as indicated by Goetghebuer (1931a: Figure 2), and we find no real differences between the paratype of *P. subcultellatum* and the German specimens of *P. cultellatum*. The figures in Rossaro (1985) of the hypopygium and pupal tergites IV and VI appear to support this conclusion. Oliver *et al.* (1990) list *P. subcultellatum* and not *P. cultellatum* in their catalog of Nearctic Chironomidae.

Polypedilum appendiculatum Goetghebuer (1939) is the only other European species in *P.* (*Uresipedilum*) and *P.* (*Polypedilum*) with several apical setae on the SVo. We have not seen specimens of this species, but it appears to be very close to *P. cultellatum*. Ashe and Cranston (1991) list this species as a probable *Polypedilum* nomen dubium.

Biology

A widely distributed and common species in Europe, but, so far, known in North America only from two specimens from California. This species has been found in lakes and small ponds (Thienemann, 1954), including subarctic lakes (Brundin, 1949). Goetghebuer (1936) associates *P. cultellatum* with stagnant eutrophic water, but Dittmar (1955) found it in the mud in the lower regions of a stream, and Lehmann (1971) found larvae on branches and stones in the weak current in the Fulda River. Lehmann concludes that it is a typical and common species of rivers.

Material Examined

—Reared Specimens

Germany, Schlitz Hutsdorf, 3 males with pupae only.

—Adult Specimens

California, 6 mi. E. Orosi, 8 July 1947, W.W. Wirth, 1 male (paratype of *P. subcultellatum* Sublette), (USNM).

Polypedilum (P.) falciforme New Species

Adult Male

Diagnosis—Small, light colored species, total length usually less than 4 mm. Antennal ratio 1.60 or less. SVo sickle-shaped, curved medially, usually thin and sharply pointed. Phallapodeme expanded at medial end. Usually 4 or fewer medial setae on each gonostylus.

Description—(n=9)

Coloration: Head, haltere, abdomen, and wing veins stramineous. Hypopygium and vertex of head tinged with light brown on one dark specimen. Pedicel and flagellum light brown. Thorax stramineous, thoracic vittae, scutellum, and ventral portion of preepisternum may be tinged with light brown. Legs stramineous, distal 1/3 of the forefemur, foretibia and foretarsal segments 1-4 usually shade darker.

Biometrics:	L total 3.50 (3.14-3.94) (8) mm	L thorax 0.98 (0.88-1.08) (8) mm
	L wing 1.86 (1.65-2.25) mm	W wing 0.56 (0.50-0.65) mm

Antenna: AR 1.45 (1.28-1.57) (8). D scape 152 (140-160) (7). L pedicel 106 (100-110) (3). L flagellomeres 1-12 429 (390-460) (7). L flagellomere 13 623 (530-720) (7).

Head: W 500 (460-520) (7). Interocular distance 111 (90-130) (7). HIR 4.50 (4.00-5.78). Vertical S 12.5 (10-16) (8). Clypeal S 17.0 (14-19) (8). Ocelli extremely small or absent. L maxillary palp segments 1-5: 37 (29-43) (8), 40 (25-47) (8), 109 (90-122) (7), 105 (95-117) (7), 183 (162-205) (6).

Thorax: Antepronotum and thoracic sensilla as in *P illinoense*. Acrostichal S 14 (1). Scutal fossal S absent. Dorsocentral S 12.4 (8-17) (7), 1 to 3 anterior setae located anteriorly in some specimens. Scutellar S in 2 regular rows, anterior row smaller; anterior row 5.3 (4-8) (7); posterior row 8.2 (6-10) (7). Prealar S 4.5 (4-6) (7).

Wing: VR 1.19 (1.14-1.27), wing unspotted but uniformly light brown in 1 specimen. Apices of R_{4+5} and M either equidistant from wing apex or R_{4+5} closer to apex than M. R_1 and R_{2+3} adjacent apically. Brachiolar S 1.1 (1-2). R_1 S 36.0 (31-43). R_{4+5} S 30.3 (26-36). Squamal S 9.1 (7-11).

Legs: L foretibial apical scale 32 (27-34). L spine on foretibial apical scale 12 (7-18) (8), tip dark in some specimens. L mesotibial apical spine 30 (25-36). L hind tibial apical spine 33 (27-41).

	Fe	T	Ta1	Ta2	Ta3
PI	843 (750-930)	523 (430-640)	1006 (910-1130) (6)	713 (650-800) (6)	506 (460-560) (6)
PII	868 (800-980) (8)	721 (640-860) (8)	460 (420-520) (8)	281 (260-310) (8)	201 (180-220) (8)
PIII	955 (830-1080)	888 (800-1080) (8)	650 (560-760) (8)	375 (340-430) (8)	312 (280-350) (8)

	Ta4	Ta5	LR	BV	SV
PI	385 (360-430) (6)	148 (140-160) (6)	1.96 (1.85-2.18) (6)	1.34 (1.29-1.39) (6)	1.34 (1.31-1.35) (6)
PII	130 (120-150) (8)	74 (70-85) (8)	0.64 (0.60-0.68) (8)	2.99 (2.76-3.12) (8)	3.48 (3.24-3.64) (8)
PIII	192 (170-210) (8)	92 (80-110) (8)	0.73 (0.67-0.76) (8)	2.54 (2.40-2.66) (8)	2.81 (2.50-3.14) (8)

Hypopygium: (Figures 69, 70). Tergite IX S 6.2 (3-8), located in long narrow middorsal area tapering posteriorly toward base of anal point; indistinctly demarcated in some specimens, Anal point straight or curved ventrally, not exceeding posterior margin of IVo, -19 (-5 to -43) (8). SVo sickle-like in appearance, curved medially from base to about 1/2 to 1/4 of its length from apex. SVo apex pointing toward area near base of anal point, approximately perpendicular to long axis of anal point. Microtrichia on base numerous but relatively small, basal S 2.4 (2-3) (8). LSVo 92 (81-104) (7). Location of distal S variable. Mean SVoR 1.61 (0.90-2.25) (8). Max L distal S 80.3 (68-90) (8). IVo similar to *P. nymphaeorum* in shape. L IVo 117 (106-131) (8). W IVo 39 (32-47) (8). Dorsal S 13.5 (11-17), usually inserted along lateral margin. Some specimens with small posterior setate papilla on IVo. Gonostylus as in *P. illinoense*, medial S 3.6 (3-4) moderately weak, distinctly shorter than SVo distal setae. Max L medial S 46 (43-52) (7). Max L

69

70

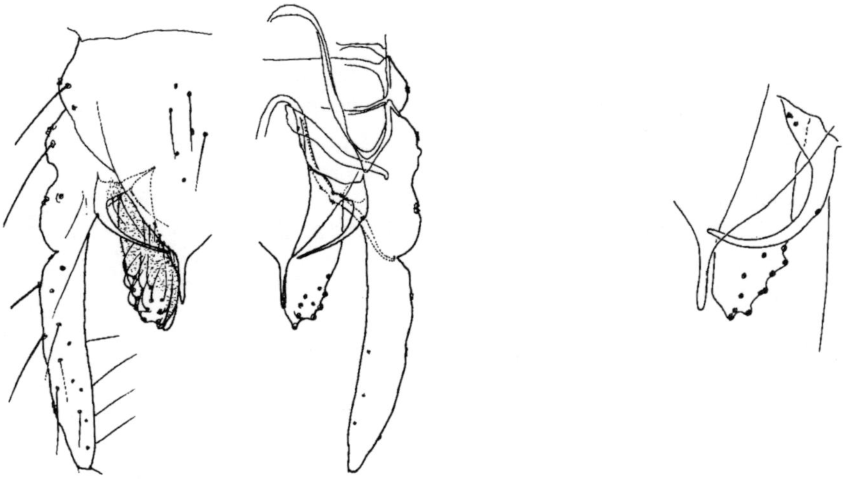

Figures 69-70. *P. falciforme* sp. nov.: 69-hypopygium; 70-variation of superior volsella.

apical S 20 (16-25) (7). L gonostylus 161 (140-185). GsR 4.86 (3.88-6.31). HR 3.39 (2.80-4.10). L posterior lobe gonocoxite 48 (45-56). HV 2.18 (1.90-2.48) (8). Phallapodeme usually bent midway along its length, medial end expanded.

Type locality: Holotype male, **Minnesota**, marshy pond, Shoreview, Ramsey Co., 45°05'N 93°08'W, 10 Aug. 1971, D.E. Maschwitz, (UMn);

Larva

Diagnosis—Very close to *P. illinoense*, distinguished only by longer third antennal segment.

Description—(n=3; all males)

Biometrics: L head capsule 193 (189-196)

Mentum: (Figure 71). W 116 (113-119), otherwise as in *P. illinoense*.

Ventromental plates: W 87 (83-88), L 39 (36-41). VmPR 2.25 (2.17-2.36). Striations (approx.) 28 (24-30). Distance between ventromental plates 55 (50-59). VmPSR 1.59 (1.44-1.77).

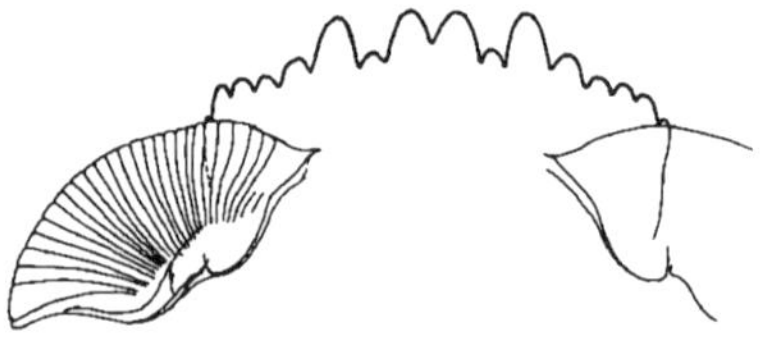

71

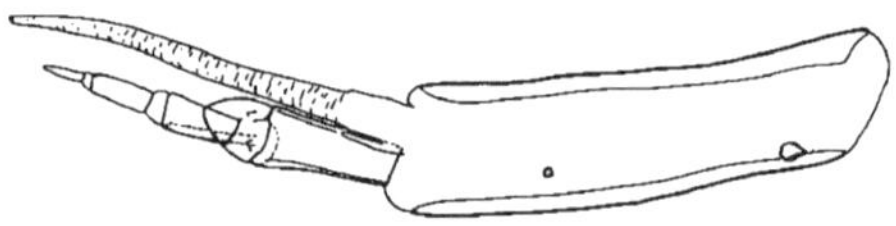

72

Figures 71-72. *P. falciforme* sp. nov.: 71-mentum; 72-antenna.

Antenna: (Figure 72). L segments 1-5: 62 (59-65) (4), 21 (21-21), 12 (12-13), 11 (10-11), 5 (5-7) (4). AR 0.78 (0.74-0.81). L primary blade 47 (41-50). L secondary blade 13 (1). L Lauterborn organs 7 (6-7), not reaching beyond middle of segment 3. L PS 12 (12-13). ROR 5.13 (4.83-5.42). Segment 1 with 1 sense pit located as in *P. illinoense*.

Labrum-epipharynx: As in *P illinoense*. PE with 3-5 teeth in lateral parts, 3 teeth in middle part.

Mandibles: L 124 (122-126). L seta subdentalis 19 (18-20) (2), otherwise as in *P. illinoense*.

Maxilla: L lacinia 46 (38-50). L segment 1 maxillary palp 21 (20-23).

Pupa

Diagnosis—Cephalothorax with tubercle located anterior to wing sheath as in *P. cultellatum* and *P. bergi* (Figure 74). *Polypedilum falciforme* can be distinguished from *P. cultellatum* by generally smaller, less distinct spinules, and from *P. bergi* by smaller number of anal taeniae.

Description—(n=3; all males)

Biometrics: L total 3.89 (3.49-4.21) mm L cephalothorax 1.05 (0.97-1.12) mm

Coloration: Typical, anal spur light brown.

Abdomen: (Figure 73). Pedes spurii B as in *P. illinoense*. Pedes spurii A weakly developed. Hooks about same size but more prominent than large anterior spinules on tergite III. Shagreen spinules small, in triangular shaped pattern on tergites II-VI, usually without fenestrations except on tergite VI. Intersegmental spinules between segments III and IV uniserial to irregularly biserial, spinule-form, pointing anteriorly; those between segments IV and V similar, except multiserial. Lateral taeniae 3 on segments V and VI; 4 on segments VII and VIII. Anal spurs with 1 main point and 0 to 2 small apical points. Anal taeniae 30 (25-34).

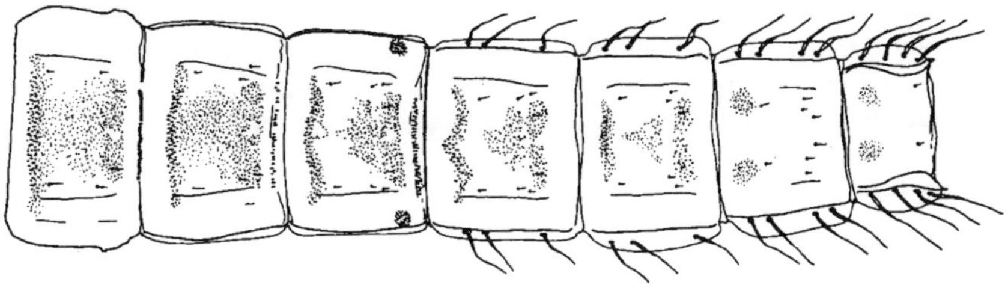

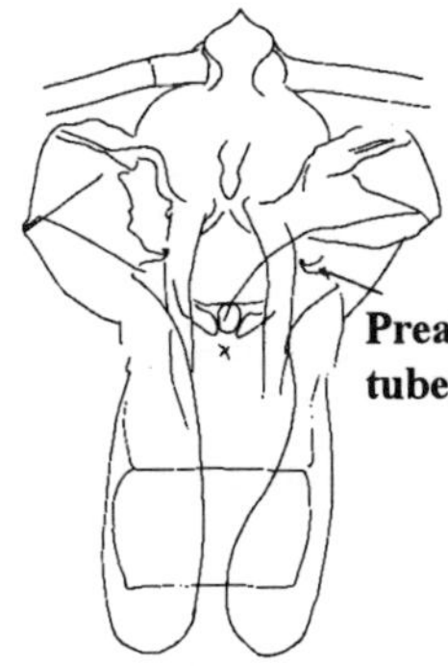

Figures 73-74. *P. falciforme* sp. nov.: 73-pupal abdomen; 74-cephalothorax showing prealar tubercle.

Remarks

The type was reared from a larva collected from cells near broken ends of *Typha* leaves or under the curled edges of *Typha* leaves. Paratypes are designated as indicated in material examined. The species is named for the sickle shaped SVo. Some specimens have a longer SVo (Figure 70). Specimens of this type can be distinguished from *P. ophioides* by the lack of setae on the medial margin beyond the basal part of the SVo.

Biology

The larva of this species appears to live in nearly the same niche as does *P. bergi*. We have reared *P. falciforme* from larvae collected in the rolled edges of *Potamogeton natans* floating leaves, and from cells in partially rotted *Typha* leaves. Larvae were usually found near the broken ends of the leaves, and some larvae were found under the curled edges of the *Typha* leaves, apparently a habitat very similar to that formed by the curled edges of *Potamogeton natans* leaves.

Material Examined

—Reared Specimens

Florida, Rayonier ditch, Flager Co., 18 March 1965, 1 male, 30 Jan. 1965, 1 male paratype, 27 Feb. 1965, 1 male paratype; Boliek's pond, Leon Co., 30 June 1964, 1 male paratype, 11 May 1965, 1 male paratype; no date, 3 males; all Florida material collected by W.M. Beck, Jr., (Beck). **Minnesota**, marshy pond, Shoreview, Ramsey Co., 45°05'N 93°08'W, 21 Aug. 1969, D.E. Maschwitz, 2 male paratypes, 10 Aug. 1971, 1 male paratype, 5 females, (UMn). Records from January through March, May, June, and August.

—Adult Specimens

Minnesota, marshy pond, Shoreview, Ramsey Co., 45°05'N 93°08'W, 22 May 1969, 5 male paratypes, 27 males, 19 June 1969, 1 male paratype, 18 July 1969, 6 male paratypes, D.E. Maschwitz, (UMn); Red Lake River near Crookston, Polk Co., 23 Aug. 1970, leg. E.F. Cook, New Jersey mosquito light trap, 1 male paratype, (UMn); Minnesota Forestry Station, Little Fork, Koochiching Co., 6 June 1970, leg. E.F. Cook, New Jersey mosquito light trap, 1 male paratype, (UMn); New London Fish Hatchery, Kandiyohi Co., 22 May 1970, leg. E.F. Cook, New Jersey mosquito light trap, 1 male paratype, (UMn). **New Jersey**, New Brunswick, 11 July 1958, W. W. Wirth, light trap, 1 male paratype, (USNM). Records from May through August.

Polypedilum (P.) fallax (Johannsen) 1905

Chironomus fallax Johannsen, 1905: 210 (female); Malloch, 1915: 435.
Chironomus (Polypedilum) fallax Johannsen, 1937: 30.
Polypedilum fallax (Johannsen) Lenz, 1941: 20; Townes, 1945: 53; McGaha, 1952: 369 (biology); Roback, 1953: 124 (biology); Paine and Gaufin, 1956: 296 (biology); Roback, 1957: 117; Beck and Beck, 1959: 93; Weber, (ed.) 1973: 28 (biology); Beck, 1977 (biology); Harper and Cloutier, 1979 (biology and emergence patterns); Soponis, 1983 (emergence pattern); Boesel, 1985: 256; Ashe and Cranston, 1991 (Palaearctic species); Caldwell *et al.*, 1997.

Adult Male

Diagnosis—Proximal three-fourths of forefemur brown, with apical half of brown portion distinctly darker than proximal half.

Description—(n=10)

Coloration: Vertex of head, pedicel, thorax, and forecoxae light brown to dark brown, scutellum shade lighter. Flagellum, maxillary palps, haltere, and abdominal segments I through V stramineous. Legs stramineous, except forefemur as noted in diagnosis, apical tarsal segments shade darker. Wing veins stramineous. Abdominal tergites VI, VII and VIII, and hypopygium light brown to brown.

Biometrics: L total 4.97 (4.06-5.59) mm; L thorax 1.29 (1.04-1.53) mm; L wing 2.55 (2.14-2.83) mm; W wing 0.73 (0.63-0.81) (9) mm

Antenna: AR 1.20 (1.10-1.26). D scape 187 (170-210) (9). L pedicel 134 (130-150) (5). L flagellomeres 1-12 649 (570-710). L flagellomere 13 781 (650-880).

Head: W 606 (530-690). Interocular distance 87 (70-130). HIR 6.97 (4.47-9.86). Vertical S 17.2 (14-20) (8). Clypeal S 27.5 (23-33). Ocelli either absent or extremely small. Maxillary palp segment 3 distinctly longer than segment 4. L maxillary palp segments 1-5: 50 (41-61), 56 (41-65), 170 (146-203), 151 (135-164), 257 (207-304).

Thorax: Dorsal edge of antepronotum moderately produced anteriorly, but not exceeding anterior margin of scutum. Antepronotal campaniform sensillum present but difficult to see on some specimens. Postpronotal depression moderately large, with many distinct campaniform sensilla. Antepronotum without setae. Humeral pit small, not visible on some specimens. Acrostichal S 19 (1), in 2 staggered rows. Scutal fossal S 3.0 (1-5) in a row roughly perpendicular to dorsocentral S (Figure 23). Dorsocentral S 16.9 (12-22), uniserial, irregularly biserial at posterior end in some specimens. Prealar S 9.0 (6-13). Scutellar S in 2 rows with anterior row of smaller setae irregular or partially double in some specimens; anterior row S 9.9 (6-16) (8); posterior row S 12.5 (9-14) (8).

Wing: VR 1.19 (1.15-1.23). Wing unspotted. Apex of R_{4+5} little to considerably closer to wing apex than apex of M. R_1 and R_{2+3} adjacent apically. Brachiolar S 1 (8). R_1 S 51.1 (41-61). R_{4+5} S 44.0 (35-53). Squamal S 20.4 (15-31) (8).

Legs: Foretibial apical scale in shape of narrow triangle, without separate distinct spine; tip usually brown (Figure 20). Total L foretibial apical scale 59 (50-68). L mesotibial apical spine 30 (25-34) (9). L hindtibial apical spine 29 (25-36) (8).

	Fe	**T1**	**Ta1**	**Ta2**	**Ta3**
PI	1166 (1000-1330)	912 (780-1030)	1368 (1160-1590) (9)	939 (810-1070) (9)	708 (610-810) (9)
PII	1374 (1190-1540)	1119 (970-1250)	653 (570-750) (9)	417 (370-480) (9)	314 (270-360) (9)
PIII	1474 (1270-1680)	1247 (1070-1410)	960 (840-1070)	555 (460-630)	441 (370-500)

	Ta4	Ta5	LR	BV	SV
PI	573 (480-670) (9)	226 (190-260) (9)	1.51 (1.43-1.68) (9)	1.42 (1.38-1.50) (9)	1.52 (1.41-1.57) (9)
PII	200 (160-230) (9)	114 (90-140) (9)	0.59 (0.57-0.60) (9)	3.01 (2.91-3.10) (9)	3.81 (3.70-3.94) (9)
PIII	277 (240-310)	133 (110-150)	0.77 (0.75-0.79)	2.64 (2.57-2.76)	2.85 (2.78-2.90)

Hypopygium: (Figure 75). Tergite IX S 10.3 (6-15) (9), located in narrow, middorsal, oval area. Anal point straight and slightly tapered, terminating near apices of IVo, 0 (-29 to 23). SVo long and usually evenly curved medially from base, usually terminating in rounded knob. Medial margin and medial part of dorsal surface of SVo base with microtrichia. Basal S 3.1 (3-4). LSVo 124 (106-133) (9). Distal seta usually placed about midway along length of SVo. Mean SVoR 0.97 (0.66-1.52) (9) (1 specimen had no distal seta). Max L distal S 130 (104-153) (9), exceeding IVo posteriorly. IVo long and narrow, nearly parallel sided or slightly narrowed apically. Apical 1/4 to 1/3 of dorsal surface free of microtrichia. Apical posteriorly directed seta present. L IVo 162 (142-179). W IVo 31 (26-41). IVo dorsal S 21.4 (18-28), moderately well developed. Gonostylus long and narrow, medial margin usually straight, lateral margin as in Figure 80. Medial S 4.7 (3-7), usually fairly evenly spaced and in straight row. Max L medial S 92 (81-108), often reaching midline. Max L apical S 52 (45-59) (4). L gonostylus 220 (191-245). GsR 3.85 (3.40-4.41). L posterior lobe gonocoxite 94 (83-108). HR 2.34 (2.16-2.72). HV 2.36 (2.06-2.96). Basal plate and medial ridge of gonocoxite distinct. Phallapodeme straight, moderately expanded midway along its length and at its medial end.

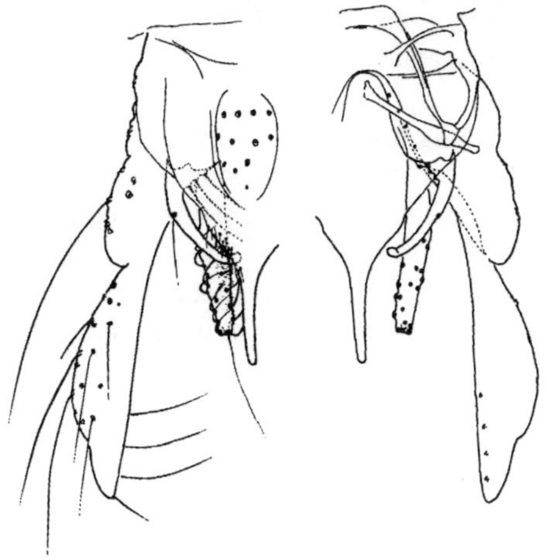

Figure 75. *P. fallax* (Johannsen), hypopygium.

Type locality: New York, Ithaca, female, (Corn).

Larva

Diagnosis—Mentum with teeth of even height as in *laetum* group. *Polypedilum fallax* distinguished from *laetum* group species by wider ventromental plates with no posterior lobe (VmPR 2.57, VmPSR 3.18), its generally larger size, and relatively smaller antenna.

Description—(n=9; 5 males, 2 females, 2 unknown).

Live color: Orange-red to red.

Eye spots: distinctly separated (Johannsen, 1937).

Biometrics: L body 7.20 (l) mm L head capsule 285 (230-369)

Mentum: (Figure 76). W 144 (113-173), with 16 teeth, median 2, first and second lateral teeth large and subequal in size, third through sixth lateral teeth smaller and subequal in size, last lateral teeth very small. In some specimens teeth appear progressively smaller from medial to penultimate teeth. Teeth uniformly dark and usually bluntly pointed.

Ventromental plates: W 149 (126-185), L 59 (50-74), without posterior lobe. VmPR 2.57 (2.45-2.84). Striations (approx.) 48, moderately distinct. Distance between ventromental plates 47 (38-56). VmPSR 3.18 (2.74-3.76).

76 77 78

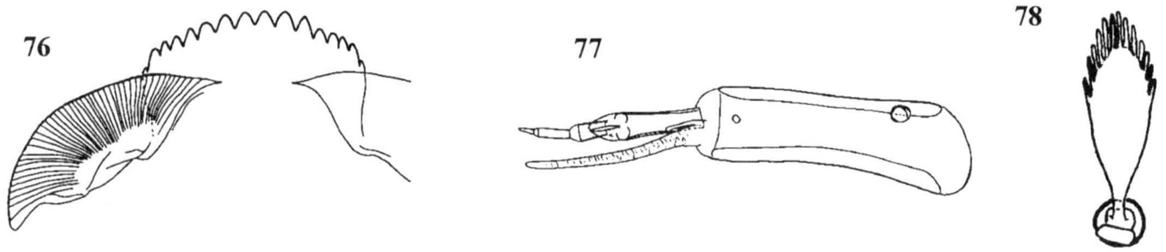

Figures 76-78. *P. fallax* (Johannsen): 76-mentum; 77-antenna; 78-labral seta I.

Antenna: (Figure 77). L segments 1-5: 65 (60-80) (7), 25 (23-28) (6), 10 (9-11) (5), 9 (9-11) (5), 5 (5-5) (4). AR 0.71 (0.67-0.75) (4). L primary blade 45 (39-50) (5), usually shorter than, or just reaching apex of segment 5. L secondary blade 10 (9-11) (2). L Lauterborn organs 9 (7-10) (4), reaching beyond middle of segment 3, sometimes to apex of this segment. Lauterborn organs very transparent and difficult to see on some specimens. Triangular structures at base of Lauterborn organs visible. L PS 9 (8-9) (3). ROR 3.95 (3.50-4.40) (6). A single large sense pit located near apex of segment 1.

Labrum-epipharynx: Labral S I very broad apically, terminating in about 14 long filamentous branches (Figure 78). S II slightly broadened apically, with fine filamentous branches along both sides of apical 1/3. S III and IV as in *P. nubeculosum*. PE in three parts, 4-6 teeth in lateral parts, 3-4 in middle part.

Mandible: L 182 (153-254). Five teeth, fifth tooth fused to medial margin. Medial margin with 3 small serrations. L seta subdentalis 29 (23-36) (7). Four branches of seta interna apparently with common origin. PMa with approximately 14 long filaments.

Maxilla: Lacinia deeply bifid, anterior branch heavier, L 68 (56-79) (8). Galea moderately well developed. Maxillary palp 3 segmented. L segment 1 28 (25-33) (8).

Pupa

Diagnosis—Fenestrations small or absent. Hooks about same size as large anterior spinules on tergite III. Shagreen spinules small, difference in size between them and anterior spinules great.

Description—(n=9; 5 males, 3 females, 1 intersex)

Biometrics: L total 5.95 (5.31-6.75) mm L cephalothorax 1.49 (1.35-1.79) (7) mm

Thoracic horn: with 8 branches (Johannsen, 1937: Figure 82).

Coloration: Typical, but generally darker with most specimens showing brownish gray shading on tergites VI, VII, and VIII.

Abdomen: (Figure 79). Pedes spurii B on segments I and II weakly to moderately developed. Pedes spurii A present. Paired row spinules on posterior lateral margins of segments V and VI. Hooks about same size as large anterior spinules on tergite III. Shagreen spinules very small compared to anterior spinules. Fenestrations absent or small. Some shagreen spinules on tergite VI grouped and arranged in rows in some specimens (all specimens of *P. calopterum*). Intersegmental spinules multiserial, distinctly larger than shagreen spinules but smaller than largest anterior spinules, pointed anteriorly. Spinules on tergites VII and VIII very small and sparse, arranged in rows. Lateral taeniae 3 on segments V and VI; 4 on segments VII and VIII. Anal spur large, with many small lateral accessory points (Figure 80). Anal taeniae 51.9 (39-62) on each anal lobe.

79

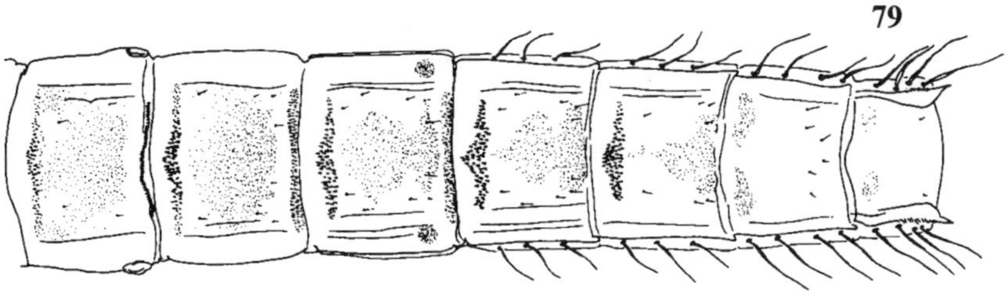

80

Figures 79-80. *P. fallax* (Johannsen): 79-pupal abdomen; 80-anal spur on tergite VIII.

Remarks

Polypedilum fallax, *P. artifer*, and *P. walleyi* are structurally indistinguishable. They are separable only on the basis of color differences. We mass-reared *P. artifer* and *P. fallax* adults from the same jar containing pieces of submerged wood, but the exuviae of the immatures could not be associated with the adult specimens. It is possible that these three species represent a single polymorphic species. However, we consider them separate species until this can be investigated.

Biology

Polypedilum fallax is a fairly common species widely distributed over the eastern half of North America. The larvae are usually found on or in submerged wood in lakes, rivers, and streams (Johannsen, 1937; Roback, 1953). We reared specimens from in and under bark on submerged logs in a small silt-laden river (Wild Rice River), and on the surface and in wood in silty to fairly clear, fast-flowing streams (Coon Creek, South Fork Root River, and Trout Creek). Also, this species was reared from larvae collected from thin-walled muddy tubes attached to the surface of rocks in the shallow water of a eutrophic lake (Deep Lake). McGaha (1952) indicates that this species is associated with the petioles of *Nuphar variegatum*. *Polypedilum fallax* is apparently facultative in its tolerance to organic pollution (Weber, 1973). Paine and Gaufin (1956) commonly found this species in both unpolluted and polluted, but recovering, portions of a stream. Beck (1977) summarized the known environmental requirements of this species. Harper and Cloutier (1979) found *P. fallax* in moderate densities and widespread throughout their study reach, with a possible preference for slower reaches over either riffles or lakes. *Polypedilum fallax* emerged throughout the summer from this Quebec stream system. In Florida, *P. fallax* emerged in the winter months from Turkey Creek (Soponis, 1983).

Material Examined

—Reared Specimens

Minnesota, Trout Creek, 7 mi. E, 2 mi. S, Chatfield, Fillmore Co., 43°50'N 92°03'W, 21 July 1971, D.E. Maschwitz, 4 males, 2 females, (UMn); Wild Rice River, Mahnomen, Mahnomen Co., 14 June 1968, D.E. Maschwitz, 2 males, 2 females, (UMn); South Fork Root River, 1.5 mi. S Houston, Houston Co., 43°44'N 91°34'W, 25 May 1969, D.E. Maschwitz, 1 female, (UMn); Deep Lake, North Oaks, Ramsey Co., 45°07'N 93°05'W, 18 May 1969, D.E. Maschwitz, 1 male, (UMn); Coon Creek, Coon Rapids, Anoka Co., 45°08'N 93°18'W, 14 May 1971, D.E. Maschwitz, 1 male, (UMn). **Ontario**, Kenora, 27 Aug. 1967, 1 female, (FRB). **Saskatchewan**, Squaw Rapid, 53°41'N 103°20'W, 6 June 1967, 1 male, 4 females, (FRB). Records from May through August.

—Adult Specimens

Approximately 80 specimens from **Connecticut**, **Florida**, **Illinois**, **Indiana**, **Maryland**, **Massachusetts**, **Minnesota**, **New York**, **Virginia**, **Wisconsin**, and **Ontario**; from April through September.

Polypedilum (*U.*) *flavum* (Johannsen) 1905

Chironomus flavus Johannsen, 1905: 225; Malloch, 1915: 474; Muttkowski, 1919: 412 (biology); Branch, 1931: 154 (description of eggs); Johannsen, 1937: 31 (description of larva and pupa).
Polypedilum flavus (Johannsen). Lenz, 1921: 158; Townes, 1945: 59, (*P. flavus* considered a synonym of *P. convictum* (Walker)); Chernovskii, 1949: 153 (figure of ventromental plate indicates that this may be a misidentification); Roback, 1953:124 (species labeled as *P.* near *illinoense* is probably *P. flavum*); Boesel, 1985: 253 (discussion of systematics and figures of hypopygium, larval and pupal characters of *P. convictum*). The following authors describe some aspects of *P. convictum* (*P. flavum* ?) biology in North America: Roback, 1957: 117; Beck and Beck, 1959: 93; Beck and Beck, 1966b: 311-312; Beck, 1977; Harper and Cloutier, 1979 (biology and emergence patterns); Soponis and Russell, 1982 (characters of early instar larvae); Soponis, 1983 (emergence patterns).
Polypedilum flavum (Johannsen). Epler, 1995: 7.94; Oyewo and Sæther, 1998: 334.
Chironomus flaviventris Johannsen, 1907: (a synonym of *P. convictum* (*P. flavum* ?), *fide* Townes, 1945: 59).
Tendipes flavus Johannsen. Bause, 1913: 115.

Adult Male

Diagnosis—Light colored species, SVo with bluntly pointed medial lobe. Apex of SVo occasionally produced into tubercle at base of distal seta. Specimens with tuberculate SVo can be distinguished from *P. obtusum* by lack of lateral expansions on anal point.

Description—(n=11)

Coloration: Flagellum, thoracic vittae, and postnotum stramineous to light brown; otherwise stramineous.

Biometrics:	L total 3.62 (2.95-3.95) (10) mm	L thorax 0.98 (0.84-1.11) (10) mm
	L wing 1.95 (1.62-2.20) (10) mm	W wing 0.57 (0.44-0.66) (9) mm

Antenna: AR 1.92 (1.71-2.21) (10). D scape 159 (140-170) (10). L pedicel 118 (110-130) (4). L flagellomeres 1-12 384 (350-420) (10). L flagellomere 13 739 (640-840) (10).

Head: W head 496 (450-530) (9). Interocular distance 113 (90-130) (9). HIR 4.41 (3.92-5.00) (9). Vertical S 12.5 (10-14) (8). Clypeal S 16.6 (14-21) (8). Ocelli small, not seen on some specimens. L maxillary palp segments 1-5: 39 (29-47) (6), 33 (25-43) (6), 139 (90-160) (8), 140 (104-158) (7), 222 (207-232) (6).

Thorax: Prothorax as in *P. illinoense*. Postpronotal depression usually without visible campaniform sensilla. Humeral pit usually visible. Acrostichal S 10 (1). Dorsocentral S 11.6 (9-15) (9). Scutal fossal S absent. Prealar S 4.8 (4-6) (10). Scutellar setae in 2 rows; setae in anterior row smaller; anterior row, S 6.1 (4-8) (10); posterior row, S 10.0 (8-11) (10).

Wing: VR 1.18 (1.13-1.19) (8). Wing unspotted. Apices of R_{4+5} and M approximately equidistant from wing apex. R_1 and R_{2+3} adjacent apically. Brachiolar S 1.1 (1-2) (8) (1 wing of 1 specimen had 2). R_1 S 37.2 (30-45) (9). R_{4+5} S 34.3 (30-41) (7). Squamal S 10.7 (7-13) (6).

Legs: L foretibial apical scale 29 (25-34) (10), without apical spine. L mesotibial apical spine 21 (18-25), L hindtibial apical spine 21 (16-27).

	Fe	**Tl**	**Ta1**	**Ta2**	**Ta3**
PI	838 (700-930)	575 (420-650)	1007 (870-1130) (9)	720 (630-800) (9)	491 (430-550) (9)
PII	937 (770-1020) (10)	795 (660-880) (10)	503 (420-560) (8)	266 (230-300) (8)	194 (160-220) (8)

PIII	965 (780-1080) (10)	912 (740-1000) (10)	708 (570-780) (8)	369 (320-400) (7)	317 (270-340) (7)
	Ta4	**Ta5**	**LR**	**BV**	**SV**
PI	379 (320-410) (9)	168 (140-190) (9)	1.71 (1.57-1.80) (10)	1.40 (1.35-1.45) (10)	1.44 (1.36-1.51) (10)
PII	128 (110-140) (8)	70 (60-80) (8)	0.62 (0.59-0.64) (8)	3.46 (3.36-3.57) (8)	3.54 (3.34-3.69) (8)
PIII	199 (170-210) (7)	100 (90-110) (7)	0.77 (0.73-0.80) (7)	2.72 (2.65-2.77) (7)	2.69 (2.60-2.76) (7)

Hypopygium: (Figures 81,82). Tergite IX S 14.4 (11-18), located in broad, middorsal, oval shaped area. Anal point curved ventrally at apex, not expanded laterally except slightly at apex. Anal point usually exceeding IVo posteriorly, 14.7 (0 to 32). SVo broad, with digitiform process at its posteromedial margin. SVo at origin of distal seta usually evenly rounded, but some specimens have distinct apical tubercle bearing distal seta, producing foot-shaped SVo (Figure 82). Microtrichia present at base of distal seta. SVo without distinct base. Setae on medial margin (basal setae) 2.0 (2-2). LSVo 85 (68-99). Mean SVoR 0.92 (0.63-1.33). Max L distal S 79 (68-101). IVo medial margin generally straight, lateral margin irregular. IVo usually broadest near midpoint. Prominent posteriorly directed seta at or near apex of IVo on ventral side. Dorsal surface free of microtrichia except proximally and along medial margin. L IVo 94 (79-106). W IVo 32 (23-41) (9). IVo dorsal S 20.2 (16-27), moderately to weakly developed. Gonostylus broad, medial margin usually slightly convex, lateral margin strongly convex. Medial S 7.8 (7-9) (10), weakly developed. Max L medial S 37 (32-47). Max L apical S 26 (18-32). L gonostylus 162 (140-176). GsR highly variable, 3.48 (2.80-4.28). HR 3.28 (2.92-3.57). L posterior lobe gonocoxite 50 (43-56). HV 2.17 (1.93-2.40) (10). Phallapodeme expanded at point of articulation and slightly expanded medially.

Type locality: **New York**, Ithaca, male.

81

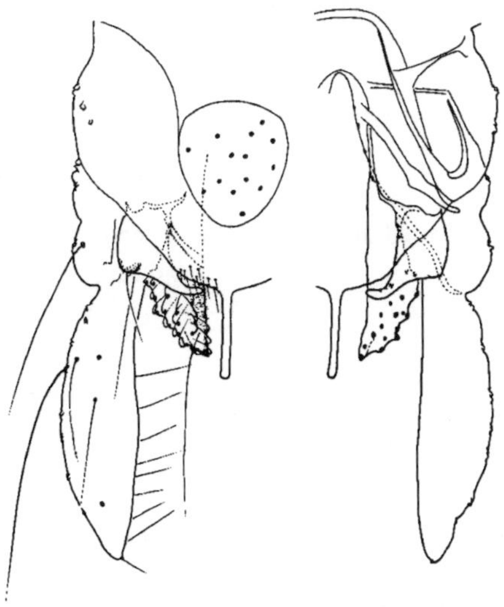

82

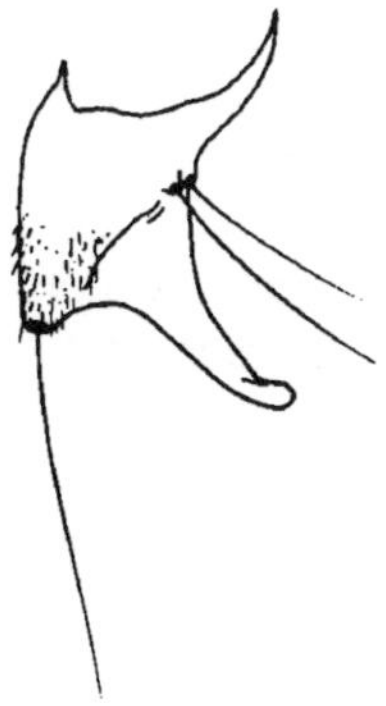

Figures 81-82. *P. flavum* (Walker): 81-hypopygium; 82-variation of superior volsella.

Larva

Diagnosis—Ventromental plate with well developed posterior lobe. Length of antennal segments 1-5: 75, 26, 8, 9, and 6. VmPR less than 2.0.

Description—(n=5, 4 males; one unknown)

Live color: Pale greenish yellow (Johannsen, 1937).

Eye spots: contiguous (Johannsen, 1937).

Biometrics: L total 6 to 7 mm (Johannsen, 1937) L head capsule 199 (189-212)

Mentum: (Figure 83). W 103 (95-110), with 16 teeth. Last lateral teeth very small, median 2 and second lateral teeth large, subequal in size. First lateral teeth small, less than half the length of median 2 teeth; fourth through seventh lateral teeth smaller, becoming progressively smaller laterally. Teeth uniformly dark, bluntly pointed.

Ventromental plates: W 80 (74-88), L 43 (38-47). VmPR 1.85 (1.75-1.95). Striations (approx.) 24, distinct except at extreme anteromedial corner. Posterior lobe well developed. Distance between ventromental plates 55 (52-61). VmPSR 1.44 (1.32-1.52).

83 84

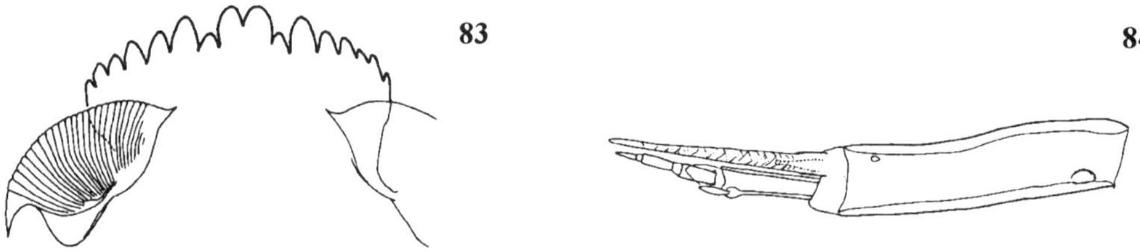

Figures 83-84. *P. flavum* (Walker): 83-mentum; 84-antenna.

Antenna: (Figure 84). L segments 1-5: 75 (73-77), 26 (25-28), 8 (7-9), 9 (8-10), 6 (5-6) (3). AR 0.65 (0.65-0.65) (3). L primary blade 49 (50-51) (3), usually exceeds apex of segment 5 but does not in some specimens. L secondary blade 12 (11-13) (4). L Lauterborn organs 6 (1), nearly reaching apex of segment 3, but extremely difficult to see, not visible in most specimens. Triangular structures visible. PS narrowed at base, lanceolate, difficult to see on some specimens. ROR 5.52 (5.36-5.64) (3). One sense pit located near apex of segment 1.

Labrum-epipharynx: Labial S I and II as in *P. illinoense*, III and IV as in *P. nubeculosum*. PE subdivided into 3 parts, with well developed teeth; lateral parts with 4.3 (4-5) teeth, middle part with 3 (3-3) teeth.

Mandible: L 125 (117-133). Teeth as in *P. illinoense*, medial margin with 3 needle-like serrations. L seta subdentalis 24 (16-27). Seta interna with 4 main branches apparently with common origin. PMa with about 10-11 (2) filaments.

Maxilla: L lacinia 49 (45-52), deeply bifid, anterior branch heavier. Galea moderately well developed. Maxillary palp 3 segmented. L segment 1 20 (18-22) (2).

Pupa

Diagnosis—Hooks on posterior margin of tergite II large and much more prominent than spinules. Intersegmental spinules between segments III and IV absent, except for possibly few scattered spinules. Spinule pattern on tergite II reduced to posterior, and narrow anterior, band of spinules with anterior band often reduced to few spinules in anterolateral corners.

Description—(n=4; all males)

Biometrics: L total 3.87 (3.73-4.01) mm L cephalothorax 1.09 (1.00-1.14) mm

Coloration: Typical.

Abdomen: (Figure 85). Pedes spurii B present on segment I, usually much reduced or absent on segment II. Pedes spurii A absent. Hooks on tergite II large and prominent. Spinules on tergite II small, reduced to anterior and posterior band. Anterior band of relatively large spinules present on tergites III through VI. Shagreen spinules on tergites III through VI concentrated on either side of midline, midline usually free of spinules. Spinule pattern with large fenestrations. Some specimens (mainly from Florida) with few spinules. Intersegmental spinules absent between segments III and IV, present between segments IV and V, pointing anteriorly. Spinules on tergites VII and VIII very small and usually arranged in rows. Lateral taeniae small; 3 on segments V and VI, 4 on segments VII and VIII. Anal spur with 1 primary point and 0 to 2 apical accessory points varying from very small to nearly as large as primary point. Anal taeniae 67.0 (43-85) on each anal lobe.

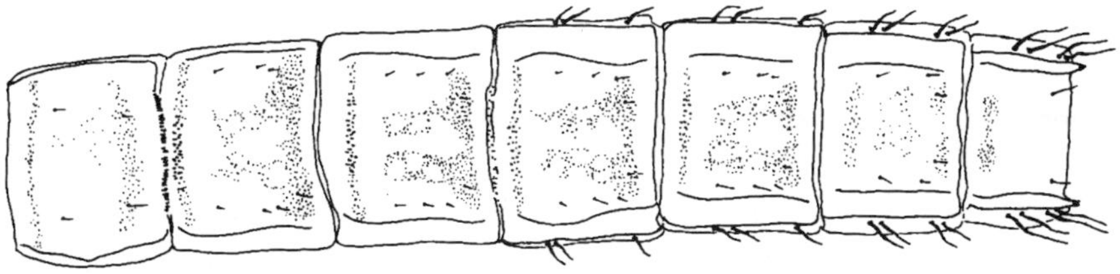

Figures 85. *P. flavum* (Walker), pupal abdomen.

Remarks

The recognition of the Nearctic *P.* (*U.*) *flavum* as a separate species from the Palearctic *P.* (*U.*) *convictum* is gaining acceptance. At the time this work was completed we accepted Townes' (1945) opinion that *P. convictum* and *P. flavum* were the same Holarctic species in spite of some expressed doubts (Beck and Beck, 1966a). However, Townes did not look at the immature stages. Rossaro (1985) points out that the mentum of European *P. convictum*, which he illustrates, have teeth of nearly even height, whereas North American specimens have teeth of uneven height (*e.g.* Johannsen, 1905: Pl. 22, Figure 1 of *C.* (*P.*) *flavus*; Soponis and Russell, 1982). The figure of a *P. convictum* mentum in Pinder and Reiss (1983) agrees with Rossaro. Also, the pupae of Rossaro's (1985) reared specimens have distinct pedes spurii A and B in contrast to the North American specimens we examined. Oyewo and Sæther (1998) treat *P. convictum* and *P. flavum* as separate species. In their review of Afrotropical species in the subgenus *P.* (*Uresipedilum*), they compare 46 adult, pupal, and larval characters for the known species in the subgenus from all zoogeographic regions, including *P. convictum* and *P. flavum* (Oyewo and Sæther, 1998). Four of 41 characters (five characters could not be determined) show clear differences between *P. convictum* and *P. flavum.* The two (of six) larval characters that differ include the differences in the mentum mentioned above, and the antennal blade exceeding the end of antennal segment five in *P. flavum* but not exceeding segment five in *P. convictum*. The condition of both characters described for *P. flavum* by Oyewo and Sæther agrees with the specimens we examined. The single (of eight) pupal character showing a difference between the two species pertains to significant differences in spinule patterns on the abdominal tergites. Again, the North American specimens we examined agree well with the description reported for *P. flavum* by Oyewo and Sæther (1998). On the other hand, they indicate that both *P. flavum* and *P. convictum* have pedes spurii A present and we considered pedes spurii A to be absent on our specimens. The single character (of 27) that separates the adults is a slightly spatulate apex of the anal point in *P. convictum* compared to a straight or tapered anal point in *P. flavum* (Oyewo and Sæther, 1998). The adult specimens we examined showed both conditions and we would not attach too much value to the ability of this character to separate the two species. It appears that *P. convictum* and *P. flavum* may be indistinguishable in the adult stage. However, on the basis of the important differences reported for the immature stages of these two species, we are treating them as separate species and *P. flavum* as a valid Nearctic species.

Johannsen's (1905) figure of the *C. flavus* hypopygium does not show the typical convictum group "boot-shaped" SVo, but he corrects this in 1937 by comparing what he saw to Goetghebuer's (1928: Figure 128) figure of a *P. convictum* hypopygium (Johannsen, 1937: 32). Boesel (1985) accepts *P. convictum* as a variable Holarctic species, whereas Epler (1995) suggests *P. convictum* and *P. flavus* (Johannsen) may be different species.

Specimens with a tubercle at the base of the apical seta on the SVo seem, at first glance, to be significantly different from those without tubercles, but there is a gradation from one condition to the other. A few specimens have one tuberculate SVo and one non-tuberculate SVo. Lehmann (1971: Figure 37) and Pinder (1974: Figure 3) show the SVo of *P.* (*U.*) *convictum* with a tubercle, whereas Edwards (1929: Figure 13L) and Goetghebuer (1928: Figure 128) show the SVo without a tubercle. Oyewo and Sæther (1998) illustrate two SVo for *P. convictum*, one with and one without a tubercle. Both the North American and European material we saw includes specimens with and without tuberculate SVo.

Biology

Polypedilum (*U.*) *flavum* is a common species and is widely distributed throughout the eastern two-thirds of North America. We reared this species from larvae collected on the surface of submerged logs in a silty river (West Fork Des Moines River) and from submerged wood in a small stream (Elm Creek). This species has been frequently collected in streams (Roback, 1957; Beck and Beck, 1966b) where it lives on rocks in riffles (Johannsen, 1905; 1937). Muttkowski (1918) reports *P. flavum* from plants in Lake Mendota. Beck (1977) summarized the environmental requirements of this species. *Polypedilum convictum* (*flavum* ?) was one of the most common species in the L'Achigan River, Quebec, where it and *P. aviceps* demonstrated a definite preference for riffle areas (Harper and Cloutier, 1979). *Polypedilum convictum* (*flavum* ?) dominated the upper portions of riffles in their study area whereas *P. aviceps* dominated the downstream sections. *Polypedilum convictum* (*flavum* ?) has a distinct bimodal emergence pattern in this river system (about 47° N latitude). The first emergence occurred the first two weeks of June and the second in mid-July (Harper and Cloutier, 1979). In contrast, Soponis (1983) found *P. convictum* (*flavum* ?) in fewer numbers relative to *P. aviceps* in Turkey Creek, and the former has an amodal, almost year-round, emergence pattern from this northern Florida stream.

Material Examined

—Reared Specimens

Florida, Swift Creek, Hamilton Co., 26 Oct. 1965, W.M Beck Jr., 13 males, (Beck); Peter's Creek, Clay Co., Nov., Jan., March, April, W.M. Beck, Jr., 8 males, (Beck); Black Creek, Clay Co., 23 Aug. 1966, 13 Oct., 1967, W.M. Beck, Jr., 7 males, (Beck); Bear Creek, Calhoun Co., 22 Oct. 1967, W.M. Beck, Jr., 1 male, (Beck); Rayonier Ditch, Flager Co., 30 April 1965, W.M. Beck, Jr., 2 males, (Beck); Rock Springs, Orange Co., 12 Feb. 1969 W.M. Beck, Jr., 1 male, (Beck). **Minnesota**, W Fork Des Moines R., Kilen Woods State Park., 6 mi. NW Jackson, Jackson Co., June 26, 1969, D.E. Maschwitz, 14 males, (UMn); Elm Creek, 4 mi. SW Anoka, Hennepin Co., 45°09'N 93°26'W, 29 May 1969, D.E. Maschwitz, 1 female (UMn). **Manitoba**, NW Creek, Heming Lake, 29 June-7 July 1967, D.D. Wiens, 1 male, (FRB). Records from January through August, October, and November.

—Adult Specimens

Over 200 specimens from: **Arizona**, **Illinois**, **Iowa**, **Kansas**, **Louisiana**, **Maryland**, **Massachusetts**, **Michigan**, **Minnesota**, **Nebraska**, **New York**, **Ohio**, **Pennsylvania**, **South Dakota**, **Texas**, **Ontario**, **Quebec**, **Germany**; collected in all months from April through September.

Polypedilum (*P.*) *illinoense* (Malloch) 1915

Chironomus illinoensis Malloch, 1915: 471; Miller, 1941: 61, 63 (biology).

Chironomus illinoensis var. *decoloratus* Malloch, 1915: 472; Townes, 1945: 57. New synonymy.

Polypedilum illinoense (Malloch). Townes, 1945: 57; Hauber, 1947: 462 (description of larva and pupa); Berg, 1950: 91 (biology, description of larva and pupa); Roback, 1957: 117 (description of pupa); Dendy and Sublette, 1959: 513; Beck and Beck, 1959: 93; Sublette, 1960: 207; Sublette, 1964: 140; Beck and Beck, 1966b: 310-313; Beck, 1977 (biology); Harper and Cloutier, 1979 (emergence patterns); Soponis and Russell, 1982 (characters of early instar larvae); Soponis, 1983 (emergence patterns); Boesel, 1985: 257 (summary of biology, figures of hypopygium, mentum, and pupal characters); Hudson *et al.*, 1990: 37 (biology and pollution tolerance); Epler, 1995: 7.94.

Adult Male

Diagnosis—Most specimens with posterior margin of tergites I-IV (II-IV in light specimens) tinged with brown; otherwise SVo long, usually only slightly curved medially from base to point just short of apex, apical one-sixth to one-twelfth bent sharply medially or ventrally.

Description—(n=10)

Coloration: Amount of brown color present highly variable, ranging from extensive tinging of brown in dark specimens to no brown color at all except on dorsal surface of scutellum and slight tinging on scutal lateral vittae in very light specimens (Malloch's var. *decoloratus*). In specimens showing brown coloration, head, pedicel, and flagellum are stramineous to brown, flagellum always distinct shade darker than former two structures. Haltere and legs stramineous to light brown, apical one-third of forefemur, tibia and tarsal segments 1-3 usually slightly darker. Thorax generally light, following areas tinged with brown: median and lateral vittae, postnotum, anepisterna II and III, and epimeron II. Abdominal tergites light, with brownish band on the posterior margin of tergites I-VI (II-IV on lighter specimens). Abdominal sternites and wing veins stramineous.

Biometrics:	L total 3.80 (3.06-4.27) (7) mm	L thorax 1.08 (0.86-1.29) (7) mm
	L wing 1.98 (1.55-2.27) mm	W wing 0.59 (0.48-0.68) mm

Antenna: AR 1.78 (1.61-1.98) (9). D scape 169 (150-180) (7). L pedicel 112 (100-130). L flagellomeres 1-12 433 (380-460) (9). L flagellomere 13 771 (710-890) (9).

Head: W 538 (500-600). Interocular distance 94 (80-120). HIR 5.72 (5.00-6.25). Vertical S 14.1 (10-17). Clypeal S 20.1(18-23). Ocelli absent or visible as extremely small spots on some specimens. L maxillary palp segments 1-5: 39 (34-45), 41 (32-50), 127 (115-135), 116 (104-128), 210 (169-234).

Thorax: Dorsal edge of antepronotum usually even with anterior most edge of scutum, appearing as flat plate on anterior scutal surface in lateral view (Figure 22). Campaniform sensillum on dorsal edge of antepronotum, postpronotal chaeticum sensillum, and humeral pit small or reduced, not visible on many specimens. Postpronotal depression usually without visible campaniform sensilla. Antepronotum without setae. Acrostichal S 16.0 (12-19) (4). Scutal fossal S 1.4 (1-2) (8). Dorsocentral S 17.8 (13-24) (9), uniserial. One campaniform sensillum located just anterior to mid point and another near posterior end of dorsocentral setal row. Prealar S 5.6 (4-9) (9). Scutellar S in 2 irregular rows, setae in anterior row smaller; anterior row S 9.4 (5-16), posterior row S 12.2 (8-16).

Wing: VR 1.20 (1.17-1.25) (9). Wing unspotted. Apices of R_{4+5} and M about equidistant from wing apex. R_1 and R_{2+3} adjacent apically. Brachiolar S 1 (7). R_1 S 45.3 (39-52). R_{4+5} S 38.1 (32-43). Squamal S 13.4 (11-17) (7).

Legs: Foretibial apical scale rounded apically, L 35 (25-43), usually with apical spine. L spine on foretibial apical scale 10 (9-14) (8), tip often dark (spine absent on one specimen). L mesotibial apical spine 30 (27-34) (9). L hindtibial apical spine 32 (27-36) (9).

	Fe	**T1**	**Ta1**	**Ta2**	**Ta3**
PI	910 (820-1000)	595 (500-680)	1042 (1000-1140)	758 (710-830)	530 (490-580)
PII	999 (860-1130) (9)	818 (700-920) (9)	489 (420-550) (9)	312 (270-350) (9)	224 (180-250) (9)
PIII	1047 (900-1170) (9)	964 (810-1090) (9)	726 (630-810) (9)	403 (350-460) (9)	337 (290-370) (9)

	Ta4	Ta5	LR	BV	SV
PI	395	166	1.76	1.39	1.43
	(360-430)	(140-190)	(1.68-1.85)	(1.32-1.46)	(1.39-1.47)
	(8)	(8)	(8)	(8)	(8)
PII	143	77	0.60	3.05	3.72
	(120-230)	(60-120)	(0.58-0.62)	(2.92-3.14)	(3.58-3.81)
	(9)	(9)	(9)	(9)	(9)
PIII	210	99	0.75	2.60	2.79
	(180-230)	(80-120)	(0.73-0.78)	(2.42-2.70)	(2.72-2.82)
	(9)	(9)	(8)	(8)	(9)

Hypopygium: (Figure 86). Tergite IX S 6.7 (5-10) (9), located in narrow, middorsal, triangular shaped area. Anal point straight or curved ventrally in some specimens, generally not exceeding IVo posteriorly, -20 (0 to -38). SVo long and angled medially from base; portion from base to just short of apex usually nearly straight or curved only slightly medially. Tip (apical one-sixth to one-twelfth) usually curved sharply medially (ventrally in some specimens). Rounded medial and dorsal surface of base with sparse, moderately weak microtrichia, basal S 3.2 (3-4). LSVo 90 (79-99). Placement of distal S variable. Mean SVoR 1.68 (1.0-2.70). Max L distal S 90 (72-113), not exceeding IVo posteriorly. IVo moderately wide, medial margin usually distinctly rounded making IVo broader and closer to midline in apical half. Entire surface covered with microtrichia except small bare area dorsally near posterolateral margin. L IVo 129 (113-140). W IVo 42 (37-54) (8). IVo dorsal S 11.1 (9-13), moderately developed; without strong apical posteriorly directed seta. Gonostylus long and narrow. Gonostylar medial S 3.4 (2-6), moderately well developed but not reaching midline; usually angled anteromedially. Max L medial S 52 (45-61). Medial S, while shorter, of about same strength as distal S on SVo. Max L apical setae 24 (16-34) (7). L gonostylus 178 (155-203). GsR 4.78 (3.90-5.80). HR 3.37 (2.76-3.76). L posterior lobe gonocoxite 53 (47-59). HV 2.19 (1.99-2.45) (8). Phallapodeme curved about midway along its length, medial end usually broadened. A group of specimens, measured separately, appears to have relatively large SVo (Figure 87). LSVo 87 (83-95). Portion of the SVo proximal to the distal seta origin is lengthened, mean SVoR 1.12 (0.64-1.50) (6). IVo generally shorter, LIVo (97-119) (6). Anal point exceeding IVo posteriorly, 10 (2-23) (5) (see Remarks).

86

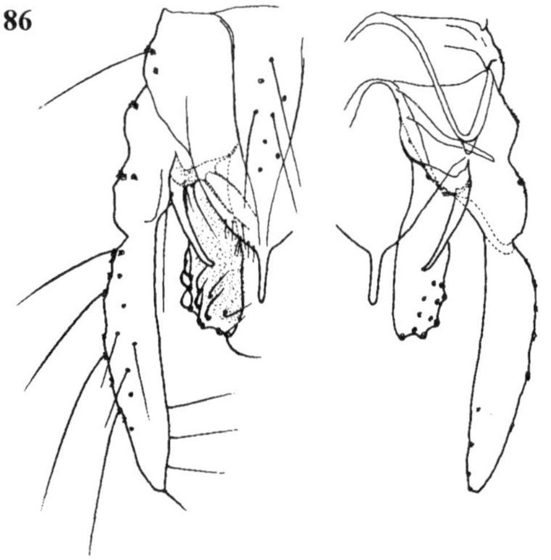

87

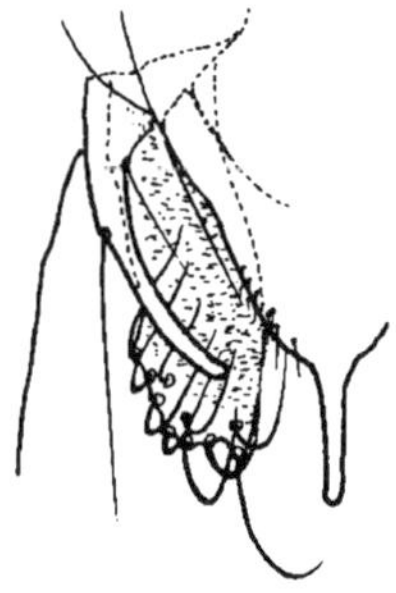

Figures 86-87. *P. illinoense* (Malloch): 86-hypopygium; 87-variation of superior volsella.

Type locality: Illinois, along bank of creek, Carbondale 23 April 1914, C.A. Hart and J.R. Malloch, taken by sweeping vegetation, male, (INH).

Larva

Diagnosis —VmPR 2.27. VmPSR 1.57. Posterior lobe of ventromental plate poorly developed, separating *illinoense* group from *P.* (*Uresipedilum*) (*convictum* group).

Description—(n=10; all males)

Live color: Greenish yellow (Hauber, 1947), pale red or orange (Berg, 1950).

Eye spots: Contiguous (Hauber, 1947).

Biometrics: L total 6 mm (Hauber, 1947; Berg, 1950) L head capsule 177 (160-187).

Mentum: (Figure 88). W 108 (101-117) (9), with 16 teeth, last lateral teeth very small, difficult to see on some specimens. Median 2 and second lateral teeth large, subequal in size. Other teeth, except last lateral, smaller and subequal in size. Teeth uniformly dark, rounded apically.

Ventromental plates: W 84 (77-90). L 37 (36-38), posterior lobe weakly developed. VmPR 2.27 (2.16-2.50). Striations (approx.) 28 (24-30), distinct except in anteromedial corner. Distance between ventromental plates 54 (50-59) (9). VmPSR 1.57 (1.48-1.67) (9).

88

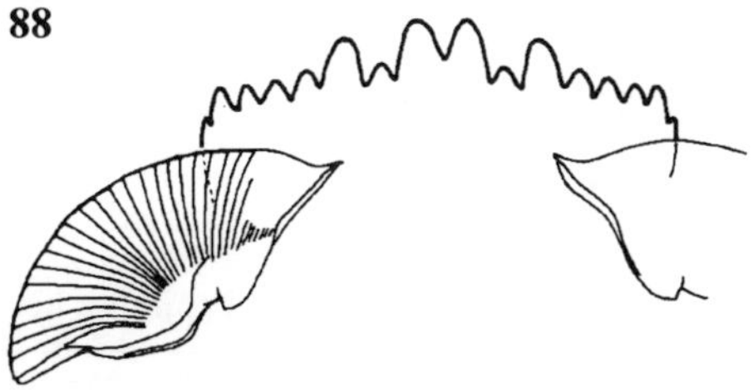

89

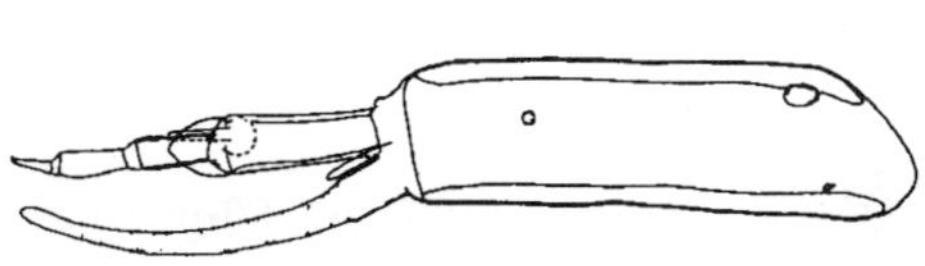

Figures 88-89. *P. illinoense* (Malloch): 88-mentum; 89-labial palp.

Antenna: (Figure 89). L segments 1-5: 58 (50-65), 20 (18-21) (11), 9 (9-10) (11), 9 (8-9), 5 (5-5). AR 0.75 (0.69-0.86) (9). L primary blade 43 (38-47), usually not exceeding apex of segment 5. L secondary blade 9 (9-10). L PS 9 (9-10) (3). L Lauterborn organ 7 (7-7) (3), arch shaped, and reaching mid part of segment 3. Triangular structures at base of Lauterborn organs visible. Ring organ located near base of segment 1. ROR 4.64 (4.36-4.85) (3). Segment 1 with 1 sense pit located about 1/3 the length of segment 1 from its apex.

Labrum-epipharynx: Labial S I moderately expanded laterally toward apex, apex filamentously branched. S II not expanded laterally and sparsely filamentously branched, appearing unbranched or branched only on one side on some specimens. S III and IV as in *P. nubeculosum*. PE in 3 parts, mid part with 3 teeth, lateral parts with 4 to 5 teeth.

Mandible: L 120 (113-124). Medial margin with 3 very sharp, needle-like serrations. L seta subdentalis 20 (18-25). Distal most branch of seta interna apparently with separate origin, other 3 branches with common origin. PMa with 3 to 6 (6) long filaments.

Maxilla: Galea slightly to moderately developed. Lacinia distinctly bifid, beak-like in appearance. L lacinia 46 (38-50) (9). Maxillary palp 3 segmented, L segment 1 19 (16-20) (7).

Pupa

Diagnosis —Shagreen spinules on tergites II-IV roughly "V" shaped.

Description—(n=10; all males)

Biometrics: L total 4.19 (3.36-4.70) mm L cephalothorax 1.14 (0.97-1.29) (9) mm

Thoracic horn: with 6 branches (Hauber, 1947).

Coloration: Typical.

Abdomen: (Figures 90, 91). Pedes spurii B moderately to well developed on segment II, weakly developed on segment I. Pedes spurii A present. Spinules on anterior margins of tergites II-VI (except laterally) distinctly larger than hooks on tergite II and intersegmental (III-IV and IV-V) spinules. Intersegmental spinules may be similar in shape to hooks, or appear as irregular bumps without points, or as normal spinules pointing posteriorly. Intersegmental spinules generally uniserial (portion of row occasionally irregularly biserial as in Figure 91). Shagreen spinules on tergites II-IV moderately small, usually in form of crude "V" pointing anteriorly; usually somewhat fenestrated. Anterolateral spinule patches on tergites VII and VIII with row spinules. Lateral taeniae 3 on segments V and VI; 4 on segments VII and VIII. Anal spur with 1 primary point and 0 to 2 smaller apical points varying from very small to nearly as large as primary point. Anal taeniae 33.2 (22-45) on each anal lobe.

90

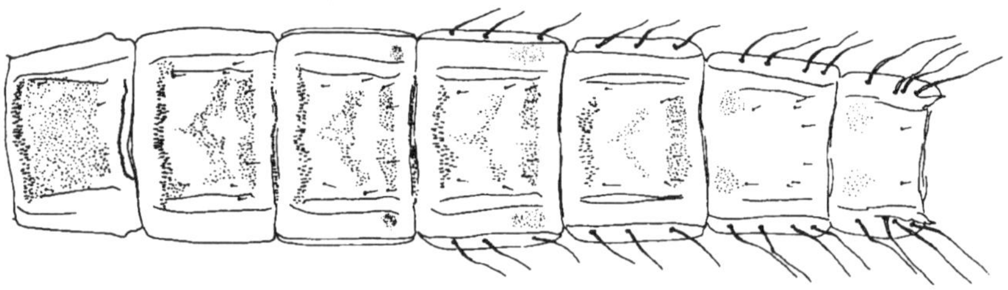

91

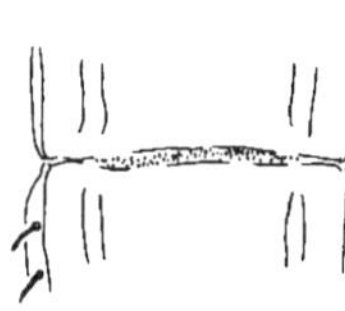

Figures 90-91. *P. illinoense* (Malloch): 90-pupal abdomen; 91-variation of intersegmental spinules between tergites IV and V.

Remarks

This species is quite variable and further splitting in the future may be justified (see the comments of Epler (1995:7.94). Also, Epler (1995) states that the antennal characters we use to separate larval *P. illinoense* from *P. ophioides* have not held up upon the examination of additional specimens. One group of specimens (from Florida, Illinois, Massachusetts, Minnesota, New York, and Ohio) has large superior volsellae and an anal point that exceeds the IVo posteriorly (Figure 87). Otherwise they are the same as the more typical *P. illinoense*. Some in this group have the typical brown coloring, but most are without it. These specimens can not be consistently separated from *P. illinoense*, and we elected to include them in this species.

Biology

Polypedilum illinoense is one of the most common Nearctic species in the genus and is distributed throughout North America. Adults have been collected during every month of the year. Larvae are found in a wide variety of habitats. We have reared specimens collected from the surface and under the bark of submerged wood in a eutrophic lake (Snail Lake, Shoreview, Minnesota); periphyton-covered rocks in a large reservoir (Lewis and Clark Lake, Yankton, South Dakota, P.L. Hudson), and from the bottom of the same reservoir in water 0.6 to 0.9 meters deep. *P. illinoense* have been found in the bottom substrate of lakes and ponds, usually eutrophic, in shallow to deep water (Miller, 1941; Tebo, 1955; Beck and Beck, 1966b), and in sluggish to fast-flowing rivers and streams (Wurtz and Roback, 1955; Paine and Gaufin, 1956; Roback, 1957; Beck and Beck, 1966b). Roback (1953) reported *P. illinoense* to be one of the most common species in the Savannah River, where he found them in algal masses, moss growing on wood, on and in submerged wood, and abundantly in floating debris mats. Harper and Cloutier (1979) describe *P. illinoense* as the only ubiquitous species in the L'Achigan River. It was found at all their stations with a possible weak preference for slower moving water. *Polypedilum illinoense* had a trimodal emergence pattern in Quebec. In Florida, *P. illinoense* emerged primarily in the winter with peaks in November and March/April (Soponis, 1983). Berg (1950) and Sublette (1957b) record the species from the lower surface of *Potamogeton natans* floating leaves, McGaha (1952) from the undersurface of *Nymphaea* leaves, and Boesel (1985) from the leaves of the water lotus (*Nelumbo lutea*). Beck (1977), Boesel (1985), and Hudson *et al.* (1990) summarize the biological information for this species and its reported tolerance to moderate pollution.

Material Examined

—Reared Specimens

Over 50 reared specimens from **Florida**, **Minnesota**, and **South Dakota**.

—Adult Specimens

Paratype male, **Illinois**, Creek Valley, Carbondale, 23 April 1914, J.M. Aldrich (USNM); plus about 370 specimens from: **Arkansas**, **Florida**, **Illinois**, **Indiana**, **Iowa**, **Louisiana**, **Maryland**, **Massachusetts**, **Michigan**, **Minnesota**, **Mississippi**, **Missouri**, **Montana**, **Nebraska**, **New Jersey**, **New York**, **North Carolina**, **North Dakota**, **Ohio**, **Oklahoma**, **South Dakota**, **Texas**, **Virginia**, **Wisconsin**, **Ontario**, and **Quebec**; from all months of the year.

Polypedilum (P.) laetum (Meigen) 1818

Chironomus laetus Meigen, 1818: 30; Mik, 1884: 201.

Polypedilum laetum (Meigen). Kieffer, 1920: 165; Goetghebuer, 1923: 126, 128 (examination of Meigen's species); Goetghebuer, 1928: 90; Hindes, 1929: 179; Edwards, 1929: 402; Goetghebuer, 1931b: 93; Kruseman, 1933: 134; Goetghebuer, 1937: 61 (figure of wing and hypopygium); Lenz, 1941: 80; Townes, 1945: 50; Jonasson, 1948: (biology); Brundin, 1949: 770 (biology, distribution); Sublette, 1960: 207; Bryce, 1960: 50 (larva); Reiss, 1968: 265 (biology); Lehmann, 1971: 501 (biology, figure of hypopygium); Sæther, 1975: 3132 (biology); Moller Pillot, 1978-1979: IV.II.14; Harper and Cloutier, 1979 (biology); Langton, 1984: 253 (figure of pupa); Rossaro, 1985 (figure of hypopygium, larval and pupal characters); Cranston *et al.*, 1989: 478 (figure of wing showing spots); Ashe and Cranston, 1991 (Palaearctic species).

Chironomus nubilipennis Meigen, 1838: 3 (*fide* Goetghebuer 1923: 127); Gimmerthal, 1845: 319.

Tendipes falciger (Kieffer), Kieffer, 1911: 21 (*fide* Goetghebuer, 1923: 126); Ashe and Cranston, 1991:304 (synonym).

Polypedilum nubilipenne (Meigen), Kieffer, 1920: 163.

Polypedilum heptastictum Kieffer, 1922a: 358 (*fide* Goetghebuer, 1937: 61 (new syn. ?)); Harnisch, 1922: 132 (biology); Goetghebuer, 1928: 90; Ashe and Cranston, 1991: 309 (*nomen dubium*).

Adult Male

Diagnosis—Dark species. Wing markings present, occasionally weak. SVo long, usually evenly curved medially, with or without posteroventral lobe on base. When present, SVo basal lobe narrow and bearing single terminal seta.

Description—(n=10)

Coloration: Pedicel, head, thorax, abdominal tergites dark brown, Legs light to dark brown, some specimens with coxae darker than rest of leg; on others, apices of femora darker than rest of leg. Flagellomeres, halteres, and palps brown, abdominal sternites light brown. Wing veins stramineous.

Biometrics:	L total 4.08 (3.74-4.42) (8) mm	L thorax 1.12 (0.99-1.28) (8) mm
	L wing 2.21 (1.85-2.45) mm	W wing 0.64 (0.56-0.71) mm

Antenna: AR 1.53 (1.31-1.82) (9). D scape 157 (140-170). L pedicel 124 (100-150) (7). L flagellomeres 1-12 449 (390-480) (9). L flagellomere 13 706 (510-780) (9).

Head: W 564 (510-640). Interocular distance 104 (90-120). HIR 5.46 (4.82-6.44). Vertical S 12.8 (10-15). Clypeal S 15.6 (9-21). Ocelli small. L maxillary palp segments 1-5: 45 (38-52), 35 (29-43), 148 (133-164), 148 (140-178) (8), 226 (203-268) (9).

Thorax: Dorsal edge of antepronotum not projecting anteriorly past anterior edge of scutum. Antepronotum appearing as flattened plate on scutum in lateral view. Postpronotal depression with visible campaniform sensilla. Antepronotum without setae. Acrostichal S 17 (1), in staggered double row. Dorsocentral S 13.8 (10-21), usually uniserial, occasionally biserial midway along row or posteriorly. Single campaniform sensillum located near posterior end of dorsocentral row. Scutal fossal S absent, but anterior most dorsocentral seta located well forward in few specimens. Prealar S 5.7 (4-8). Scutellar S 17.9 (14-23), in 2 rows, setae in anterior row smaller.

Wing: VR 1.08 (1.01-1.12). Wing with five spots located as follows: just distal of r-m cross vein, just distal to fork of M_{3+4} and Cu_1, at apices of M_{3+4} and Cu_1, and fifth spot on anal vein near wing margin. Apices of R_{4+5} and M about equidistant from wing apex. R_1 and R_{2+3} usually separated near their apices. Brachiolar S 1. R_1 S 42.7 (40-46) (9). R_{4+5} S 36.3 (27-47). Squamal S 13.9 (10-17) (7).

Legs: Foretibial apical scale with apical spine; spine with darkened tip in some specimens. L scale 42 (36-52). L spine 11 (7-20) (9). L mesotibial apical spine 16 (14-20). L hindtibial apical spine 18 (14-23) (9).

	Fe	T1	Ta1	Ta2	Ta3
PI	897 (770-970)	750 (640-820)	1044 (900-1090)	625 (520-670)	482 (410-510)
PII	1015 (860-1130)	838 (720-920)	528 (470-580)	295 (250-330)	219 (190-250)
PIII	1094 (950-1190)	976 (830-1060)	768 (570-860)	417 (360-460)	338 (290-370)

	Ta4	Ta5	LR	BV	SV
PI	366 (320-390) (9)	158 (140-190) (9)	1.41 (1.20-1.58)	1.66 (1.63-1.75) (9)	1.58 (1.50-1.77) (9)
PII	135 (120-150)	87 (80-100)	0.63 (0.58-0.66)	3.24 (3.13-3.38)	3.52 (3.36-3.78)
PIII	199 (180-220)	105 (90-120)	0.78 (0.69-0.83)	2.68 (2.52-2.77)	2.71 (2.61-3.12)

Hypopygium: (Figures 92, 93, 94). Tergite IX S 6.6 (5-9), located in middorsal lighter, oval-shaped area. Apex of tergite IX usually rounded at base of anal point; tergite IX truncate or even concave on either side of anal point base in some specimens. Anal point tapered slightly from base to apex, length variable in relation to length of IVo, 5 (-16 to 32). SVo of 2 types: first with narrow lobe off posterior portion of SVo base (shown in dotted line ventral to SVo in Figure 92), which bears 1 or more setae, but only 1 terminal seta. Proximal portion of base with 1 seta. Distal portion of SVo usually evenly curved medially toward anal point, nearly reaching anal point just proximal to middle of anal point. Second SVo type with base rounded posteroventrally, without lobe (Figures 93, 94). Total basal S 2.6 (2-4). LSVo 96 (88-115). Distal seta variable in position. Mean SVoR 0.75 (0.44-1.41). Max L distal S 69 (56-83). IVo angled medially so that apices closer together than bases; apex often truncate, and with posteriorly directed seta. IVo dorsal S 11.2 (9-14), moderately strong. L IVo 129 (114-140). W IVo 30 (25-42). Gonostylus small; medial S 4.8 (4-7), moderately strong, usually not reaching midline. Max L apical S 41 (20-54) (7). L gonostylus 127 (119-135). GsR 3.06 (2.68-3.80). Gonocoxite with large posterior lobe, L posterior lobe 66 (61-70). HR 1.94 (1.71-2.19). HV 3.20 (2.90-3.49). Microtrichia on basal plate and along medial margin of IVo very large. Phallapodeme similar to *P. nubeculosum* with medial end expanded but indistinct.

Type locality: **France**, female, (? Paris museum).

Larva

Diagnosis—Mentum teeth of even height. Separated from *fallax* group species by shape of ventromental plate and relatively narrow labral S I. Primary antennal blade reaching only to apex of segment 4.

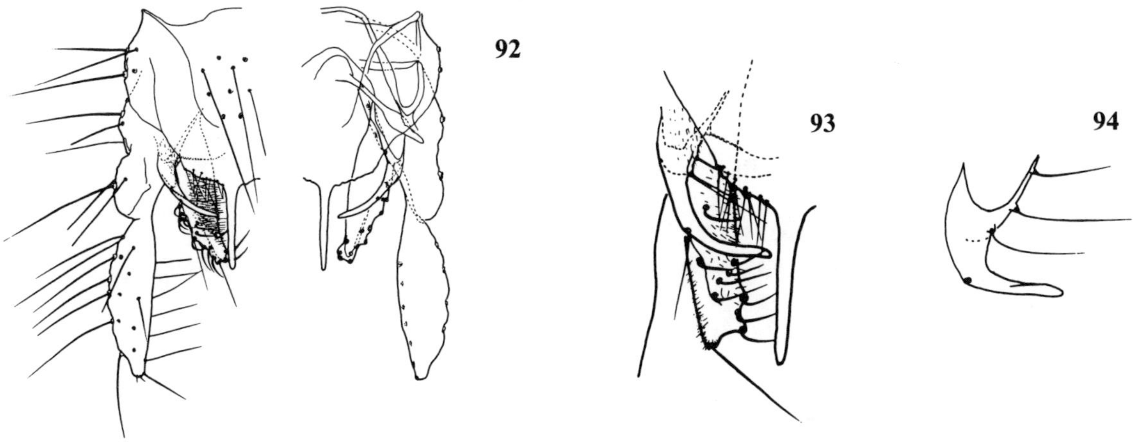

Figures 92-94. *P. laetum* (Meigen): 92-hypopygium; 93-variation of the superior volsella; 94-variation of the superior volsella.

Description—(n=4; 1 male, 1 female, 2 unknown)

Biometrics: L head capsule: 191 (178-212)

Mentum: (Figure 95). W 98 (90-113), with 16 teeth (14 if very small last lateral teeth are not counted), sharply pointed in specimens seen, all teeth uniformly dark and generally even in height. Medial 6 teeth distinctly larger than remaining teeth, last lateral tooth extremely small.

Ventromental plates: W 108 (95-122). L 48 (47-50), posterior lobe moderately developed. VmPR 2.31 (1.91-2.60). Striations as in *P. nubeculosum*. Distance between ventromental plates 56 (45-68) (3). VmPSR 2.09 (1.80-2.30).

Antenna: (Figure 96). L segments 1-5: 71 (58-80), 22 (15-27), 16 (11-19), 19 (13-22), 7 (6-8) (3). AR 0.88 (0.76-0.93). L primary blade 62 (52-67) (3), reaching only to apex of segment 4. L secondary blade 9 (8-9) (3). L PS on segment 2 10 (8-12) (3). L Lauterborn organ 10 (10-10) (2), small and rounded, difficult to see. Ring organ on segment 1 located close to base, ROR 4.44 (3.88-4.82). Segment 1 apparently with 2 sense pits.

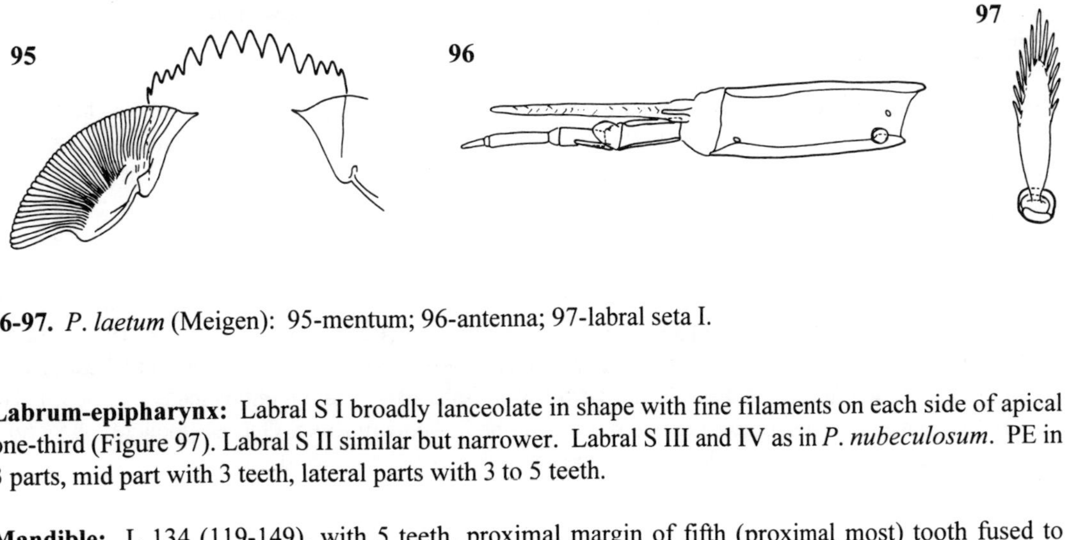

Figures 96-97. *P. laetum* (Meigen): 95-mentum; 96-antenna; 97-labral seta I.

Labrum-epipharynx: Labral S I broadly lanceolate in shape with fine filaments on each side of apical one-third (Figure 97). Labral S II similar but narrower. Labral S III and IV as in *P. nubeculosum*. PE in 3 parts, mid part with 3 teeth, lateral parts with 3 to 5 teeth.

Mandible: L 134 (119-149), with 5 teeth, proximal margin of fifth (proximal most) tooth fused to anteromedial margin of mandible, same margin with 2 serrations. L seta subdentalis 29 (25-34) (3), tapered to fine point at apex. Seta interna with 4 main branches, PMa as in *P. nubeculosum*.

Maxilla: As in *P. nubeculosum*. L lacinia 47 (45-50) (3). Maxillary palp with 3 segments, L segment 1 26 (20-29) (3).

Pupa

Diagnosis—Shagreen spinules relatively large, about same size as hooks on posterior margin of tergite II. Shagreen pattern covering most of tergites II and III.

Description—(n=3; 2 males, 1 female)

Biometrics: L total: 4.69 (4.42-4.87) mm L cephalothorax: 1.17 (1.14-1.24) mm

Coloration: Typical.

Abdomen: (Figure 98). Pedes spurii B on segment I large, those on segment II small. Pedes spurii A on segment IV moderately developed. Hooks on tergite II smaller than spinules on anterior margins of tergites II-VI. Intersegmental spinules multiserial, smaller than shagreen spinules, pointing anteriorly. Shagreen spinules large, spinule pattern filling most of tergites II through IV, pattern somewhat triangular shaped on tergite V, with large fenestrations on all tergites. Tergites VII and VIII with sparse anterolateral patches of very small row spinules. Lateral taeniae 3 on segments V and VI; 4 on segments VII and VIII. Anal spurs with 1 large apical point, with 2 to 3 smaller apical points and many small lateral accessory points. Anal taeniae 25.8 (19-33) (2) on each anal lobe.

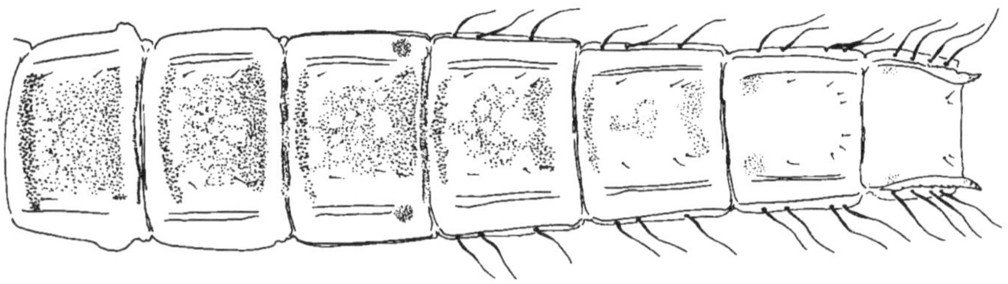

Figure 98. *P. laetum* (Meigen), pupal abdomen.

Remarks

We accept *T. falciger*, *C. nubilipennis*, and *P. heptastictum* as synonyms of *P. laetum* based on the work of Goetghebuer. Sæther (1975) speculates that the species called *P. laetum* in North America may not be the same the European *P. laetum*. Rossaro's (1985: 18) figures show considerable variation in the SVo, and he suggests that *P. laetum* and *P. castornama* sp. nov. may be the same species.

Among the specimens examined are two rather distinct types which correspond fairly closely to the two hypopygial figures of *P. laetum* in Townes (1945: 182, 202). Three of our measured specimens, two western and one from Ithaca, New York, have a truncate ninth tergite and lack the posteroventral lobe at the base of the SVo. The remaining measured specimens generally have a rounded ninth tergite and a distinct posteroventral lobe on the base of the SVo.

Biology

Larva of *P. laetum* are typically found in eutrophic lakes (Brundin, 1949; Sæther, 1975). Larvae of this species have been found in bottom mud in the littoral zone of a moderately eutrophic lake (Reiss, 1968). Lehmann (1971) found *P. laetum* in sandy mud in riffle areas of the River Fulda and on lightly silted moss and stones. Lindegaard-Peterson (1972) also found it in the middle and lower reaches of a northern European stream where it builds loose sand tubes. Harper and Cloutier (1979) report *P. laetum* only from two stations in the L'Achigan River, Quebec (moderate flow rates) in late August and in very small numbers. Epler (1995) reported it from northern Florida.

Material Examined

—Reared Specimens

Manitoba, Winnipeg, River, Pinawa 30 May 1967, R. Ireland, bottom mud, 2 meters deep, 1 male, 1 female (FRB). **Ontario**, Kenora, 23 Aug. 1967, 1 male, (FRB). **Germany**, Fulda River, 24 July 1967, bottom mud, (UMn).

—Immature Specimens

Minnesota, Mississippi River, Minneapolis, 26 April 1973, 8 May 1973, 2 larvae, bottom sediments, (UMn).

—Adult Specimens

Approximately 140 specimens from **California, Colorado, Maryland, Massachusetts, Minnesota, Nevada, New York, Oregon, Utah, Washington, Wisconsin, Wyoming, Alberta, New Brunswick, Ontario**; from May through October.

Polypedilum (P.) nubeculosum (Meigen) 1804

Chironomus nubeculosus Meigen, 1804: 18; Meigen, 1818: 29; Zetterstedt, 1850: 3569; Walker, 1856: 162; Schiner, 1864: 598; v.d.Wulp, 1877: 269; Johannsen, 1908: 278; Goetghebuer, 1921: 139 (figure of hypopygium); Johannsen, 1926: 277; Edwards, 1929: 402.

Polypedilum nubeculosum (Meigen). Kieffer, 1920: 166; Goetghebuer, 1928: 89; Hindes, 1929: 179; Krusemann, 1933: 132 (figure of wing, hypopygium); Goetghebuer, 1937; 62: Meuche, 1939 (biology); Lenz, 1941: 18; Stuart, 1941 (biology, misidentification ?); Brundin, 1949: 770; Chernovskii, 1949: 121, 153 (figure of larva); Thienemann, 1954: many pages; Palmén, 1955: 18 (biology of adult); Dittmar, 1955: 476 (biology); Mundie, 1957: 195 (biology); Palmén, 1958a, 1958b (adult emergence); Bryce, 1960: 50; Morgan and Waddell, 1961: (biology); Lenz, 1962: 241; Hamilton, 1965: 121; Reiss, 1968: 265, 288 (biology); Sandberg, 1969: 142; (biology); Lehmann, 1971: 501, 546 (biology, figure of hypopygium); Hamilton, 1971: 260 (biology); Sæther, 1975: 3132 (biology); Moller Pillot, 1978-1979: IV.II.18 (*nubeculosum* group); Pinder and Reiss, 1983: 410 (figures of larval characters); Langton, 1984: 253 (figure of pupa); Rossaro, 1985: 19 (figures of hypopygium and larval and pupal characters); Boesel, 1985: 253 (distribution in U.S. and biology, figure of antepronotum); Ashe and Cranston, 1991 (Palaearctic species); Sasa and Kikuchi, 1995 (figure of hypopygium).

Chironomus enotatus Walker, 1856: 170 (*fide* Goetghebuer, 1928: 89); Kieffer, 1906: 17.

Polypedilum emarginatum Kieffer, 1913: 16 (*fide* Edwards, 1929: 402), (designated type species for genus *Polypedilum*); Kieffer, 1920: 161; Lenz, 1921: 157, 158, 162; Goetghebuer, 1928: 91.

Polypedilum hirtimanus Kieffer, 1915: 82 (reference not seen), (*fide* Goetghebuer, 1931a: 92); Kieffer, 1920: 160; Lenz, 1921: 157, 158, 162; Goetghebuer, 1928: 92; Ashe and Cranston, 1991: 309 (*nomen dubium*).

Polypedilum ciliatimanus Kieffer, 1915: 82 (reference not seen), (*fide* Goetghebuer, 1928: 89), (*nomen dubium*); Kieffer, 1920: 161; Lenz, 1921: 162; Harnisch, 1922: 131, 132, 142.

Polypedilum obliteratum Kieffer, 1920: 164 (*fide* Goetghebuer, 1937: 62), (*nomen dubium*); Harnisch, 1922: 129, 142 (biology); Ashe and Cranston, 1991: 309 (*nomen dubium*).

Polypedilum obliteratum var. *opacum* Kieffer, 1920: 164 (*nomen dubium*).

Polypedilum obliteratum var. *majus* Kieffer, 1922b: 92 (*nomen dubium*); Lenz, 1941: 19.

Polypedilum hirtimanus var. *scutellare* Kieffer, 1922b: 91 (*nomen dubium*); Lenz, 1921: 157, 158.

Polypedilum ciliatimanus var. unnamed Kieffer, 1915 ? (reference not seen), (*nomen dubium*); Kieffer, 1920: 161; Kieffer, 1921a: 285

Adult Male

Diagnosis—Among the largest Nearctic species of *Polypedilum*. Uniformly dark brown in color, wing with faint spots. Dorsal edge of antepronotum anterior to scutum; antepronotum with setae. SVo long, portion distal to origin of distal seta curved medially. Gonostyli large.

Description—(n=10)

Coloration: pedicel, thorax, abdominal tergites dark brown. Head, legs, haltere, flagellum brown. Abdominal sternites shade lighter than tergites. Wing veins stramineous.

Biometrics:	L total 6.37 (5.26-7.16) mm	L thorax 1.60 (1.34-1.75) mm
	L wing 3.18 (2.81-3.63) (9) mm	W wing 0.87 (0.76-1.02) (9) mm

Antenna: (Figure 2). AR 2.14 (1.94-2.34). D scape 203 (190-220) (9), with expanded flange ventrolaterally. L pedicel 151 (140-170) (7). L flagellomeres 1-12 459 (420-480). L flagellomere 13 978 (900-1090).

Head: (Figure 6). W 736 (700-800) (9). Interocular distance 146 (110-160) (9). HIR 5.12 (4.24-6.36) (9). Vertical S 19 (14-24) (9). Clypeal S 38 (31-45) (9). Coronal apodeme V-shaped, very prominent in cleared specimens. Tentorium expanded midway along its length, exact shape varying with orientation of specimen on slide. Tentorium continuing past posterior tentorial pit, extension sometimes appearing as small knob at posterodorsal end of tentorium. Ocelli extremely small, not visible on some specimens. L maxillary palp segments 1-5: 56 (47-61), 42 (38-50), 196 (169-227), 196 (182-221), 274 (207-326).

Thorax: (Figure 7). Dorsal edge of antepronotum projecting anteriorly beyond anterior margin of scutum. In dorsal view antepronotum appears as 2 rounded lobes on either side of medial commissure; campaniform sensillum present on each lobe. Distinct depression with numerous campaniform sensilla located at posterior angle of postpronotum. Chaeticum sensillum and 1 to 2 campaniform sensilla located on anterodorsal portion of postpronotum. Antepronotum with 7 (6-11) (8) setae. Acrostichal S 25 (1), in alternating double row. Dorsocentral S 37 (32-42) (8), multiserial. Several campaniform sensilla associated with dorsocentral setae row. Scutal fossal S absent. Humeral pit varies from large to small. Prealar S 14 (12-17) (9). Scutellar S 36 (30-45) (9), in random arrangement except for posterior row forming straight line; anterior setae smaller.

Wing: (Figure 3). VR 1.09 (1.06-1.13) (9). Wing membrane without setae. Marginal setae strong along costa, weaker with long and short setae intermixed along posterior wing margin, becoming generally longer on anal lobe. Alula bare. Spots on wing membrane as described by Goetghebuer (1936) and other European authors. Wing spots very weak on specimens examined. Spot just posterior to r-m cross vein usually most distinct. Weak spot usually visible posterior to fork of M_{3+4} and Cu_1 and occasionally continuing along margins of these veins posteriorly. Wing membrane may be weakly darkened at apex of R_{4+5} and from that point anteriorly along R_{4+5} to apex of R_1. Apices of R_{4+5} and M about equidistant from apex. R_{2+3} separated from R_1 at their apices. Wing veins with campaniform sensilla as follows: brachiolum with three groups of sensilla, distal group of about 13, medial group of three, and proximal group of about 10; 3 on subcosta, 1 on Fr, and 1 on R_1 near Fr. Brachiolar S 2 (1-3) (8). R_1 S 62 (57-66) (9). R_{4+5} S 46 (41-49). Squamal S 27 (22-42) (8).

Legs: Foretibial apical scale with or without apical spine (Figure 4). Scale with large seta usually placed midway between apex and base plus 2 or 3 large setae and single small seta located near base of scale. L scale 58 (50-72) (9). L spine when present 8 (5-11) (3). L mesotibial apical spine 20 (11-25) (9). L hindtibial spine 25 (11-27).

	Fe	**T1**	**Ta1**	**Ta2**	**Ta3**
PI	1126 (1000-1230)	998 (870-1140)	1340 (1090-1470)	794 (650-880)	685 (560-760)
PII	1326 (1170-1490) (9)	1189 (1080-1320) (9)	715 (600-800) (8)	413 (330-460) (8)	339 (290-380) (8)
PIII	1382 (1240-1530)	1444 (1290-1600)	1083 (980-1200) (8)	585 (520-640) (8)	446 (400-490) (8)

	Ta4	**Ta5**	**LR**	**BV**	**SV**
PI	512 (420-560)	230 (200-250)	1.35 (1.16-1.61)	1.57 (1.48-1.70)	1.59 (1.41-1.83)
PII	207 (190-230) (8)	140 (120-150) (8)	0.59 (0.55-0.61) (8)	2.99 (2.88-3.18) (8)	3.59 (3.45-3.78) (8)
PIII	265 (230-290) (8)	165 (140-180) (8)	0.73 (0.71-0.76) (8)	2.74 (2.71-2.78) (8)	2.68 (2.60-2.74) (8)

Hypopygium: (Figures 9, 99). All surfaces except anal point, SVo, and small part of dorsal surface of IVo, covered with microtrichia. Tergite IX S 15 (10-22), located in mid-dorsal, oval shaped area. Posterior margin of tergite IX with 13.8 (11-15) (9) setae on each side of anal point base. Anal point long and straight but not exceeding IVo posteriorly, -19(-5 to -50). SVo long and nearly straight to point of distal setal origin where SVo curves medially. SVo base small, bearing few microtrichia along medial margin. Basal S 1.9 (1-2). LSVo 122 (106-133). Distal seta apical in position, mean SVoR 0.62 (0.42-0.92). Max L distal S 132 (101-151). IVo long, broadened and rounded apically. L IVo 209 (188-232). W IVo 44 (38-51). IVo dorsal S 23 (15-27), moderately developed; posteriorly directed seta present. Gonostylus large, lateral margin rounded, apex often rounded. Medial S 8.0 (5-10), moderately well developed, usually curved anteriorly. Max L medial S 118 (95-144). Max L apical S 59 (52-70) (4). L gonostylus 255 (218-281). GsR. 3.16 (2.43-3.47). Gonocoxite with large anteroventral portion and distinct posterior lobe (basistyle). L posterior lobe 98 (90-106). Basal plate distinct with many large microtrichia on its surface, large microtrichia continuing posteriorly along basal-medial margin of IVo. 4.8 (4-7) (9) large setae along ventral margin of basal plate. HR 2.60 (2.31-2.78). HV 2.53 (2.30-2.82). Sternapodeme with broad lateral portion and narrow transverse portion. Phallapodeme expanded just medially from point of articulation, and at medial end.

99

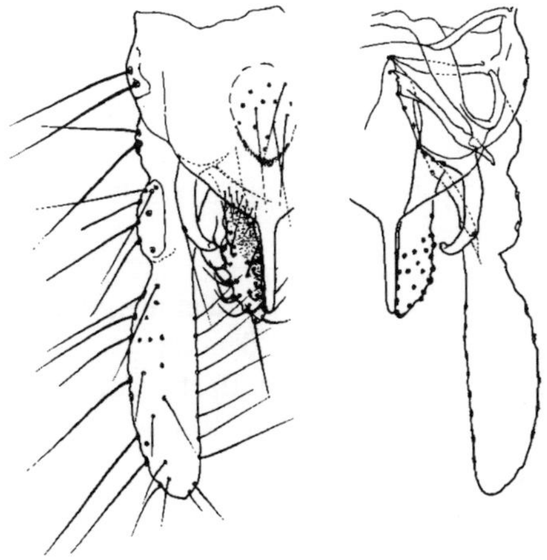

100

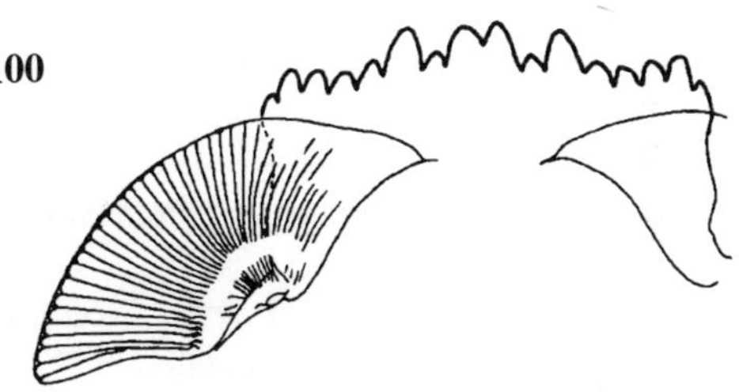

Figures 99-100. *P. nubeculosum* (Meigen): 99-hypopygium; 100-mentum.

Type locality: France, Autumn, male, (? Paris Museum).

Larva

Diagnosis—Antennal segment 3 longer than 4; Lauterborn organ not reaching beyond middle of antennal segment 3. Labral S I broader apically than in *P. laetum* group species or in *P. trigonus*. VmPSR greater than 2.5.

Description—(n=3; 1 male, 2 unknown)

Live color: red (Hamilton, 1965).

Eye spots: separate, on vertical line (Hamilton, 1965).

Biometrics: L total 11-12 mm (2); 14 mm (Hamilton, 1965) L head capsule 242 (236-245)

Mentum: (Figure 100). W 161 (160-162) (2), with 16 teeth, median 2 teeth and second laterals large, extending beyond others; remaining teeth subequal in size with penultimate laterals somewhat larger and last laterals smaller. Tips of teeth rounded, uniformly dark.

Ventromental plates: W 153 (137-167). L 63 (56-68), posterior lobe weakly developed. VmPR 2.43 (2.14-2.58). Striations (approx.) 35, distinct except at medial margin. Distance between ventromental plates 50 (47-52) (2). VmPSR 2.99 (2.74-3.24) (2).

Antenna: (Figure 101). L segments 1-5: 79 (65-88), 27 (25-29), 20 (16-23), 15 (14-16) (2), 6 (5-7) (2). AR 0.83 (0.79-0.90). L primary blade 57 (56-58) (2), reaching nearly to apex of segment 5. L secondary blade 7 (1). PS absent. L Lauterborn organ 11 (11-12) (2), arch shaped, not reaching middle of segment 3. Triangular structures at base of Lauterborn organs visible. Ring organ on segment 1 located close to base. ROR 4.36 (4.0-4.83). Segment 1 with two sense pits.

101

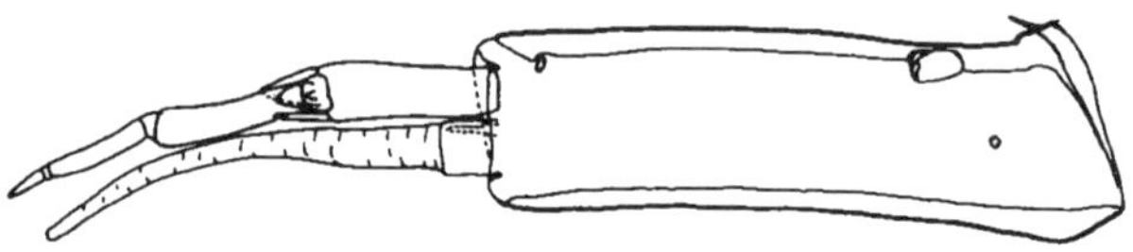

102

Figures 101-102. *P. nubeculosum* (Meigen): 101-antenna; 102-labral seta I.

Labrum-epipharynx: Labral S I broadened apically, apex filamentously branched (Figure 102). Labral S II slightly broadened apically, lateral margins with filamentous branches. Labral S III very fine, setate. Labral S IV short, rod-like. LL with many long filamentous branches. PE in three parts, mid part with 3 teeth, lateral parts, 4 to 5 teeth. Premandibles bifid.

Mandible: L 163 (140-180), with 5 teeth, most proximal very small. Anteromedial margin with three large serrations. L seta subdentalis 28 (27-29) (2), unbranched and rod-like. Seta interna with 4 main branches, 3 of these apparently with common base, distal most with separate base. Each branch subdivided into many fine filaments. Ring organ located at base near posterior mandibular seta. PMa with 5 to 6 filaments.

Maxilla: Galea slightly developed. Lacinia prominent, bifid, projecting medially. L lacinia 68 (1). Maxillary palp with 3 segments. L segment 1 28 (21-36) (2). Apex of segment 1 with (approximately) 10 sensilla.

Pupa

Diagnosis —Shagreen pattern sparse. Intersegmental spinules between segments III and IV, and IV and V prominent and multiserial. Anal spurs with many small lateral accessory points. Size large, total length usually greater than 7 mm.

Description—(n=4; 1 male, 3 females)

Biometrics: L total: 7.28 (7.0-7.6) mm. 6.4-7.2 mm (Hamilton, 1965)
L cephalothorax: 1.64 mm (1)

Thoracic horn: With 8 branches (Lenz, 1941)

Coloration: Typical (see generic description).

Abdomen: (Figure 103). Pedes spurii B moderately to well developed. Pedes spurii A well developed. Hooks on tergite II prominent and larger than spinules on anterior margins of tergites II-VI. Intersegmental spinules multiserial, pointing anteriorly, approximately same size as anterior spinules on tergites II-VI. Shagreen spinules on tergites II-V often forming central roughly triangular shaped patch, with large fenestrations often merging posteriorly with smaller posterolateral circular patches. Tergites VII and VIII with anterolateral patches of small row spinules. Lateral taeniae, 3 on segments V and VI, 4 on segments

VII and VIII. Anal spurs with 1 large primary point, many lateral accessory points, and 1 to 2 apical accessory points varying from very small to nearly as large as primary point (Figure 104). Anal taeniae 35.3 (32-41) on each anal lobe, taeniae up to 1 mm long.

103

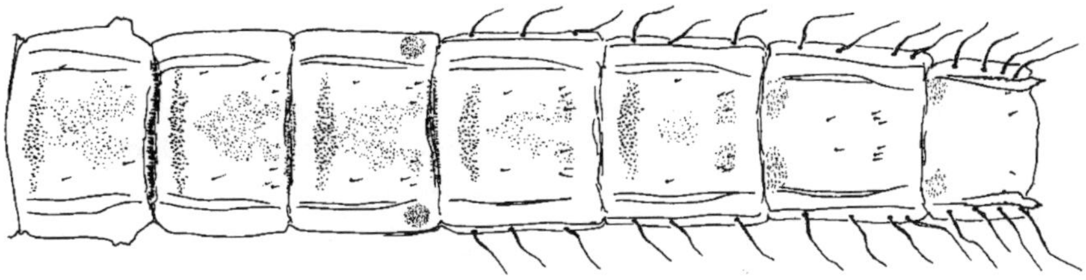

104

Figures 103-104. *P. nubeculosum* (Meigen): 103-pupal abdomen; 104-anal spur on tergite VIII.

Remarks

In recording *C. enotatus*, *P. emarginatum*, *P. hirtimanus*, and *P. ciliatimanus* as synonyms of *P. nubeculosum*, we accept Goetghebuer's work because we have not seen the types of any of these species.

There has been a good deal of confusion in the literature as to whether *P. nubeculosum* has a tarsal beard and distinct or weak wing spots. The descriptions of *C. enotatus* and *P. emarginatum* fit the concept of *P. nubeculosum* as described above (*i. e.*, no beard, weak wing spots). However, *P. hirtimanus* is described as having no wing spots and a bearded foretarsus, whereas *P. ciliatimanus* is described as having spotted wings. Johannsen (1926) said that his specimens agreed with those identified by Schiner (Vienna) and Staeger (Copenhagen) as *P. nubeculosum* in having no beard but definite wing spots. Edward's (1929) description indicates weak wing spots and a short beard; Goetghebuer's (1937) description is similar. Townes (1945) mentions weak wing spots. Rossaro (1985) indicates that some specimens have wing spots and others do not.

Goetghebuer (1937) questionably placed *P. obliteratum* in synonymy with *P. nubeculosum*. The species, described by Kieffer in a key (Kieffer, 1920), seems to fit *P. nubeculosum*; however, Lenz (1941) placed this species and *P. obliteratum* var. *majus* in a list of species in *P.* (*Uresipedilum*) (*convictum* group) based on the immatures, thus recognizing them as valid. Lenz (1941 and 1962) also lists the three Kieffer synonyms *P. ciliatimanus*, *P. emarginatum*, and *P. hirtimanus* under *P. nubeculosum*, apparently recognizing them as varieties. It seems best at this time to accept *C. enotatus*, *P. emarginatum*, *P. hirtimanus*, and *P. ciliatimanus* as synonyms *fide* Goetghebuer, and to designate the following as *nomina dubia*: *P. hirtimanus* var. *scutellare* Kieffer (1922b), *P. ciliatimanus* var. (unnamed) Kieffer (1915?), *P. obliteratum* Kieffer (1920), and *P. obliteratum* varieties *opacum* Kieffer (1920), and *majus* Kieffer (1922b).

Biology

References to the biology of *P. nubeculosum* are common in the literature. In general this species is a mud-bottom dweller preferring the shallow water of eutrophic lakes and rivers (Meuche, 1939; Mundie, 1957; Hamilton, 1965; Reiss, 1968; Sandberg, 1969; Dittmar, 1955; Lehmann, 1971).

Polypedilum nubeculosum was the dominant bottom benthic organism in a small shallow lake (surface area 10 ha, mean depth 1.5 m) in western Ontario but was less common in larger deeper lakes in the same area (Hamilton, 1971). However, the species is adaptable to a wide range of trophic conditions (Sæther, 1975). Reiss (1968) found it at 106 meters in Lake Constance, and Palmén (1955) reports it living in brackish water. Both Mundie (1957) and Reiss (1968) show three generations per year for this species in Europe, and Palmén's (1955; 1958a; 1958b) work on diel periodicity shows this species' peak emergence around midnight in June and between 9:00 and 11:00 PM in August. Hindes (1929) reported seeing *P. nubeculosum* swarming at Lake Windermere. Boesel (1985) summarizes some of the biological information for this species.

Material Examined

—Reared Specimens

Manitoba, Heming Lake, 23 June 1967, leg. A P. Wiens, depth l m, 1 male, (FRBW).

—Immature Specimens

British Columbia, Marion Lake, 29 April 1964, 1 larva; no date, 1 larva, 1 pupa (CNC). **Ontario** Brewer Lake, Algonquin Park, 14 May 1966, D. W. Webb, 4 pupae, (INH).

—Adult Specimens

Indiana, Big Cedar Lake, 20 June 1965, 7 males, (JES); Crooked Lake, 27 June 1961, 6 males, 30 June 1965, 3 males (JES). **Michigan**, Manistee Co., 17 July 1948, R. R Dreisbach, 1 male, (USNM). **Minnesota**, Itasca State Park Biological Station, Clearwater Co., 22 May 1960, 1 male, 7-18 June 1967, 15 males, 10 June 1968, 3 males, 15 June 1968, 1 male, at light, D. E. Maschwitz, (UMn); pond, Shoreview, Ramsey Co., 45°05'N 93°08'W, 26 May 1969, D. E. Maschwitz, 1 male, (UMn); Lower La Salle Lake, Hubbard Co., 10 June 1970, leg. E. F. Cook, New Jersey mosquito light trap, (UMn); 2 mi. N. Hovland, Cook Co., 27 July 1970, leg. E. F. Cook, New Jersey mosquito light trap, (UMn). **New York**, Chautauqua Lake, July 1957, E. C. Bay, at light, 1 male, (USNM); Wellesley Island, Jefferson Co., 22 Aug. 1963, L. L. Pechumen, 8 males, (JES); Lake George, 21 June 1920, 3 males, (Corn). **Manitoba**, Heming Lake, 19 June 1967, A. P. Wiens, 1 male, (FRBW). **Ontario**, Brewer Lake, Algonquin Park, 14 May 1966; D. W. Webb, 3 males, (INH); Sara-jewun Lake, Wildlife Research Station, Algonquin Park, 3 June 1966, J. Martin, at light, 3 males, (CNC). **Germany**, Bodensee, 19 Aug. 1962, 2 males, (JES); Plon, Eutiner Str. 34, at light, 26 Aug. 1964, 1 male, (UMn); Kossau, Zucht, 1 male, (UMn); 3 males with no date (UMn). Records from May through August.

Polypedilum (*P.*) *nymphaeorum* New Species

Adult Male

Diagnosis—SVo narrow and tapered to point apically. SVo generally shorter and smaller than in other species in the *illinoense* group. SVo distal seta usually placed about mid way between base and apex. IVo usually bearing prominent, posteroventral, setate papilla. Phallapodeme long, relatively straight and not expanded at medial end.

Description—(n=7)

Coloration: Generally light. Head, pedicel, haltere, wing veins, abdomen stramineous. Flagellum light brown. Thorax stramineous to light brown, dorsal and lateral vittae, dorsal part of scutellum, and preepisternum tinged with brown on some specimens (those from West Palm Beach, Florida). Legs stramineous, distal one-fourth of the forefemur, foretibia, and foretarsal segments 1-4 often shade darker.

Biometrics:	L total 3.56 (2.95-4.48) mm	L thorax 1.02 (0.94-1.12) mm
	L wing 1.81 (1.57-1.99) (6) mm	W wing 0.55 (0.51-0.61) mm

Antenna: AR 1.50 (1.44-1.55) (3). D scape 160 (150-180). L pedicel 100 (1). L flagellomeres 1-12 460 (450-470) (3).

Head: W 524 (500-560). Interocular distance 97 (50-120). HIR 5.40 (4.25-10.00). Vertical S 11.3 (9-14). Clypeal S 18.3 (14-21). Ocelli very small, not visible in most specimens. L maxillary palp segments 1-5: 38 (29-45) (6), 39 (34-45) (6), 91 (83-101) (6), 95 (90-104) (6), 156 (126-176) (5).

Thorax: Antepronotum as in *P. illinoense*. One scutal fossal S present in some specimens (these setae may be anterior most dorsocentral S). Dorsocentral S 16.1 (13-21), uniserial, usually 1 or 2 setae placed anteriorly. Prealar S 4.6 (4-6). Scutellar S in 2 regular rows, anterior row smaller; anterior row 6.8 (4-10); posterior row 8.6 (7-10).

Wing: VR 1.19 (1.17-1.21) (6). Wing unspotted. Apices of R_{4+5} and M about equidistant from wing apex. R_1 and R_{2+3} adjacent to nearly adjacent apically. Brachiolar S 1. R_1 S 40.7 (31-48). R_{4+5} S 27.7 (21-33). Squamal S 9.6 (7-14).

Legs: L foretibial apical scale 35 (32-43) (6). L spine on foretibial scale 11 (7-14) (6), tip dark on some specimens (Figure 19). L mesotibial apical spine 23 (18-27) (4). L hindtibial apical spine 26 (25-27) (4).

	Fe	T1	Ta1	Ta2	Ta3
PI	825 (740-920) (6)	563 (480-610) (6)	1037 (1020-1060) (3)	703 (690-720) (3)	503 (490-510) (3)
PII	918 (830-1010) (6)	767 (690-820) (6)	473 (450-500) (4)	283 (270-300) (4)	208 (190-230) (4)
PIII	978 (890-1090) (6)	913 (820-990) (6)	695 (660-760) (4)	393 (370-420) (4)	328 (310-350) (4)

	Ta4	Ta5	LR	BV	SV
PI	390 (380-400) (3)	143 (140-150) (3)	1.77 (1.72-1.81) (3)	1.43 (1.42-1.46) (3)	1.39 (1.35-1.42) (3)
PII	135 (120-150) (4)	88 (80-90) (4)	0.74 (0.71-0.77) (4)	2.66 (2.55-2.78) (4)	2.86 (2.75-2.95) (4)
PIII	188 (170-200) (4)	100 (90-110) (4)	0.59 (0.57-0.61) (4)	3.16 (3.04-3.36) (4)	3.74 (3.62-3.83) (4)

Hypopygium: (Figure 105). Tergite IX S 10.3 (8-12), located in middorsal elliptical shaped area tapering posteriorly toward base of anal point. Anal point straight or curved slightly ventrally, usually not exceeding IVo posteriorly, -13.5 (5 to -54) (8). SVo tapered to point apically, usually evenly curved medially, extent of curve varying from slight to moderate. Medial margin and dorsal surface of basal portion with sparse and very weak microtrichia (may not be visible on some specimens). Basal S 3.4 (3-4) (8). LSVo 81 (77-86) (4). Mean SVoR 1.18 (0.81-1.77). Max L distal S 115 (97-135) (8), usually extending posteriorly well past apex of IVo. IVo similar to *P. illinoense* except narrower, apical ½ to 1/3 of medial margin usually expanded and rounded medially. L IVo 114 (92-126) (8). W IVo 39 (32-47). Dorsal S 17.4 (14-21), moderately well developed. Posteroventral setate papilla usually present. Gonostylus as in *P. illinoense*; medial S 3.5 (2-5), weak and short. Max L medial S 50 (45-61). Apical S about one-third to one-half as long as medial setae. L gonostylus 165 (155-173) (6). GsR 4.79 (4.17-5.43) (6). HR 3.44 (3.14-3.85) (6). L posterior lobe gonocoxite 47 (45-52) (8). HV 2.25 (1.85-2.56) (5). Phallapodeme long, relatively straight, medial end not expanded.

105

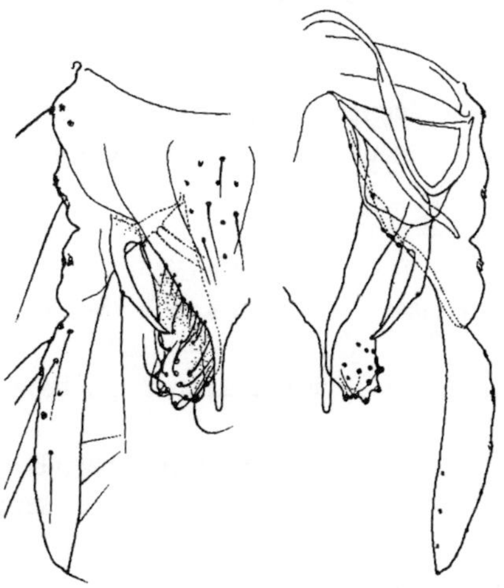

Figure 105. *P. nymphaeorum* sp. nov., hypopygium.

Type locality: Holotype male, **Minnesota**, Deep Lake, North Oaks, Ramsey Co., 45°07'N 93°05'W, 10 Aug. 1971, D.E. Maschwitz (UMn).

Larva

Diagnosis—Very similar to *P. illinoense* and *P. falciforme*. May not be consistently distinguishable from these species, except possibly by greater VmPSR. Antennal segment 3 longer than 4 should separate it from *P. angulum*.

Description—(n=5; all males)

Biometrics: L head capsule 196 (191-200)

Mentum: (Figure 106). W 104 (95-110), otherwise as in *P. illinoense*.

Ventromental plates: W 91 (81-99), L 41 (36-45). VmPR 2.24 (2.21-2.28). Striations (approx.) 27 (26-30). Distance between ventromental plates 50 (43-54). VmPSR 1.85 (1.75-1.95).

Antenna: (Figure 107). L segments 1-5: 55 (50-60), 19 (19-20) (4), 11 (9-12) (4), 10 (9-11) (4), 5 (5-5) (4). AR 0.80 (0.75-0.86) (4). L primary blade 43 (1). L secondary blade 10 (9-11) (2). L Lauterborn organs 9 (7-10) (4), reaching middle or just beyond middle of segment 3. L PS 10 (7-13) (3). ROR 4.21 (3.87-4.67). Segment 1 with 1 sense pit located as in *P. illinoense*.

106

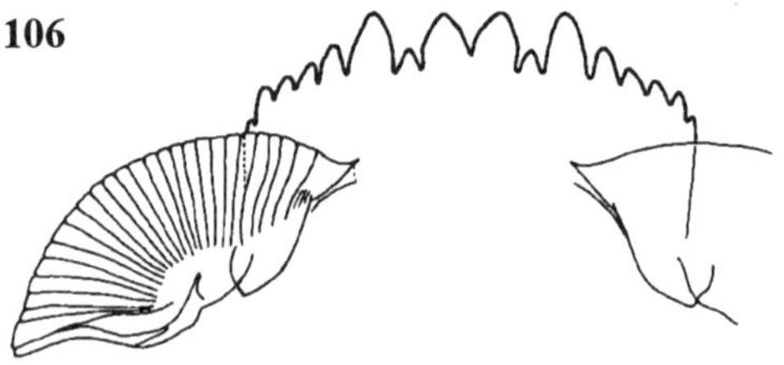

107

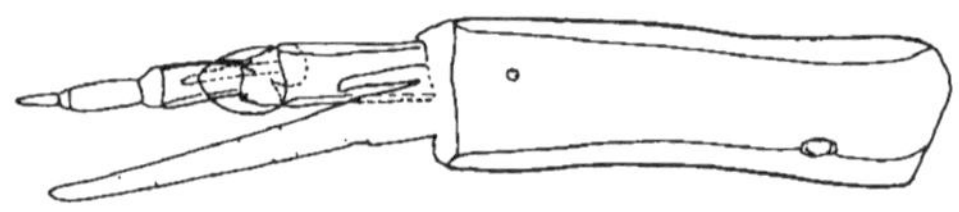

Figures 106-107. *P. nymphaeorum* sp. nov.: 106-mentum; 107-antenna.

Labrum-epipharynx: As in *P. illinoense*. PE with 3-4 teeth in lateral parts, 2-3 teeth in middle part.

Mandible: L 122 (115-128). L seta subdentalis 19 (14-23), otherwise as in *P. illinoense*.

Maxilla: L lacinia 47 (43-54). L segment 1 maxillary palp 19 (18-20) (3).

Pupa

Diagnosis—Shagreen pattern roughly triangular in shape on tergites II-V. Anterior rows of large spinules on tergites III-V about equal in size to hooks on posterior margin of tergite II. Anterior row of spinules on tergite II small.

Description—(n=5; all males)

Biometrics: L total 4.15 (3.88-4.48) mm L cephalothorax 1.15 (1.06-1.19) mm

Coloration: Typical, anal spur light brown to brown.

Abdomen: (Figure 108). Pedes spurii B on segments I and II as in *P. illinoense*. Pedes spurii A weakly to moderately developed. Hooks about same size or slightly larger than large anterior spinules on tergite III. Spinules in anterior row on tergite II distinctly smaller than those in anterior rows on tergites III-VI. Shagreen spinules small, in triangular shaped pattern on tergites II-V. Fenestrations large but few in number on tergites IV-VI. Intersegmental spinules between segments III and IV uniserial or irregularly uniserial, in form of dull spines or normal spinules, directed anteriorly or posteriorly. Intersegmental spinules between segments IV and V irregularly multiserial, spinule-form, and directed anteriorly or posteriorly. Intersegmental spinules nearly as large as anterior spinules on tergites III-VI. Lateral taeniae 3 on segments V and VI; 4 on segments VII and VIII. Anal spurs with 1 main point and 1 to 5 apical accessory points. Anal taeniae 38.0 (34-44).

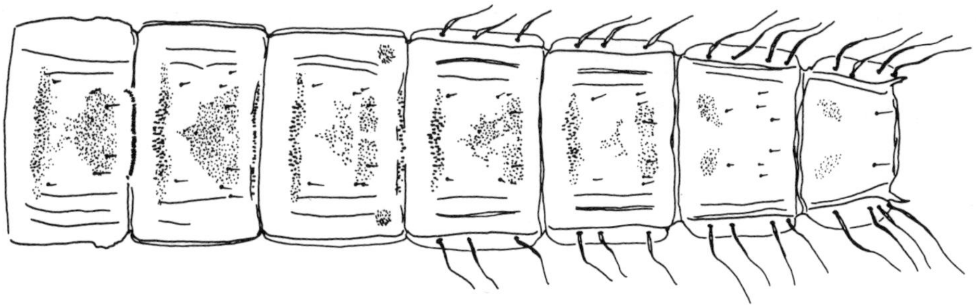

Figure 108. *P. nymphaeorum* sp. nov., pupal abdomen.

Remarks

The type was reared from a larva collected from under surface of *Nymphaea* leaf in 0.6 to 0.9 meters of water in a eutrophic lake. Paratypes are designated as indicated in material examined. The species is named for its association with the water lily *Nymphaea*.

Biology

We have reared this species from larvae collected from the bottom surface of floating *Nymphaea odorata* leaves in the shallow water of two very eutrophic lakes.

Material Examined

—Reared Specimens

Florida, Lake Susan, Lake Co., 14 July 1966, 1 male paratype, 10 Aug. 1966, 1 male paratype, 10 Sept. 1966, 1 male paratype; Lake Jackson, Leon Co., 2 Nov. 1968, 1 male; Boliek's pond, Leon Co., 11 May 1965, 1 male; no data, 3 males; all Florida material collected by W.M. Beck, Jr., (Beck). **Massachusetts**, Sturbridge, 9 Aug. 1966, 1 male paratype, (CNC). **Minnesota**, Deep Lake, North Oaks, Ramsey Co., 45°07'N 93°05'W, 10 Aug. 1971, D.E. Maschwitz, 1 male and 1 female paratypes, (UMn); Snail Lake, Shoreview, Ramsey Co., 45°04'N 93°08'W, 9 June 1971, D.E. Maschwitz, 1 female paratype, (UMn). Records from May through September and November.

—Adult Specimens

Florida, Morrison Field, West Palm Beach, 10 Dec. 1942, 2 male paratypes, 20 Sept. 1942, 1 male paratype, (JES); Chantilly Aires, Gainesville, Alachua Co., 8 May 1967, F.S. Blanton, 1 male paratype, (USNM). **Illinois**, Brussels, 26 Sept. 1939, Frisson and Ross, 1 male, (INH). **Kentucky**, Campbell Co., 27 Aug. 1969, W.T. Mason, sticky trap, 2 male paratypes, (JES). **Louisiana**, Kilbourne, 5 April 1947, W.W. Wirth, light trap, 1 male, (USNM). **Michigan**, Huron River, Washtenaw Co., C.O. Berg, 16 Aug. 1942, 2 male paratypes, (USNM). **Minnesota**, Deep Lake, North Oaks, Ramsey Co., 45°07'N 93°05'W, 24 June 1969, D.E. Maschwitz, 5 male paratypes (UMn); Lower LaSalle Lake, Hubbard Co., 29 July 1970, leg. E. F. Cook, New Jersey mosquito light trap, 1 male paratype, (UMn); Gray Eagle, 20 Aug. 1940, 1 male, (JES); 3 mi. N. Stillwater, Washington Co., 9 July 1969, D.E. Maschwitz, 1 male paratype, (UMn). **New Jersey**, New Brunswick, 11 July 1958, W.W. Wirth, light trap, 2 males, (USNM). **Ontario**, Costello Lake, Algonquin Park, 21 Aug. 1964, D.W. Webb, 1 male paratype, 4 males, (INH). Records from April through September and December.

Polypedilum (*U.*) *obtusum* Townes 1945

Polypedilum obtusum Townes, 1945: 60; Sublette, 1957a: 37-40 (larval, pupal description); Beck and Beck, 1959: 93; Beck, 1977 (biology); Boesel, 1985: 253 (*P. convictum*, new synonymy); Oliver *et al.*, 1990: 51 (listing of synonymy).

Adult Male

Diagnosis—This species is distinguished from *P. flavum* by presence of lateral expansions on anal point from base to apex. Tubercle at base of distal seta on SVo is well developed. This species usually has more dorsal setae on IVo and more gonostylar medial setae than *P. flavum*.

Description—(n=10)

Coloration: As in *P. flavum*.

Biometrics: L total 3.63 (2.56-4.06) mm; L thorax 1.07 (0.81-1.21) mm; L wing 1.94 (1.43-2.29) mm; W wing 0.58 (0.41-0.71) mm

Antenna: AR 1.92 (1.67-2.05). D scape 157 (120-170) (9). L pedicel 120 (1). L flagellomeres 1-12 403 (360-460). L flagellomere 13 772 (600-860).

Head: W head 519 (490-530) (7). Interocular distance 111 (100-130) (7). HIR 4.70 (3.77-5.30) (7). Vertical S 12.7 (11-14) (8). Clypeal S 16.2 (13-20) (9). Ocelli small, not visible on many specimens. L maxillary palp 1-5: 39 (27-45) (6), 33 (27-41) (6), 130 (115-142) (6), 133 (117-146), 219 (164-254) (6).

Thorax: Prothorax as in *P. illinoense*. Postpronotal depression small, without visible campaniform sensilla. Humeral pit present. Dorsocentral S 13.7 (9-17) (8). Scutal fossal S absent. Prealar S 4.8 (4-6) (8). Scutellar setae in 2 rows, setae in anterior row smaller; anterior row S 5.9 (4-10) (8); posterior row S 11.6 (9-15) (8).

Wing: VR 1.18 (1.15-1.21). Wing unspotted. Apices of R_{4+5} and M approximately equidistant from wing apex. R_1 and R_{2+3} adjacent to slightly separated apically. Brachiolar S 1.0 (1-1) (7). R_1 S 39.1 (30-47) (8). R_{4+5} S 31.5 (24-37) (8). Squamal S 10.3 (8-12) (8).

Legs: L foretibial apical scale 29 (25-32), without apical spine. L mesotibial apical spine 19 (18-20) (4). L hindtibial apical spine 23 (23-25) (4).

	Fe	T1	Ta1	Ta2	Ta3
PI	860 (700-970)	599 (450-680)	1011 (810-1130) (7)	724 (570-830) (7)	504 (410-580) (7)
PII	954 (750-1080)	819 (620-930)	488 (390-570) (9)	264 (220-300) (9)	193 (150-220) (9)
PIII	979 (780-1140)	930 (710-1040)	717 (550-860)	366 (290-430) (9)	304 (240-360) (9)

	Ta4	Ta5	LR	BV	SV
PI	380 (320-430) (7)	155 (140-180) (6)	1.73 (1.66-1.81) (7)	1.39 (1.36-1.43) (6)	1.41 (1.38-1.45) (7)

PII	121 (90-140) (9)	70 (60-80) (9)	0.60 (0.58-0.63) (9)	3.46 (3.38-3.55) (9)	3.61 (3.51-3.76) (9)
PIII	181 (150-210) (9)	88 (70-100) (9)	0.78 (0.75-0.82) (9)	2.79 (2.72-2.89) (9)	2.65 (2.56-2.71) (9)

Hypopygium: (Figures 109, 110, 111). Tergite IX S 17.9 (11-23), located in very broad, middorsal, oval shaped area. Anal point dorsal surface expanded laterally from base to rounded apex. Lateral expansions varying from relatively narrow with margins semiparallel from base to apex (Figure 109) to broad with widest part near base (Figures 110, 111). W of anal point at base 23 (16-34). Apex of anal point curved ventrally. Anal point usually exceeding IVo posteriorly, 15.2 (-23 to 43). SVo as in *P. flavum* except that apical tubercle at base of distal seta usually larger. Microtrichia and basal setae as in *P. flavum*. Basal S 2 (2-2). LSVo 103 (86-110). Mean SVoR 0.98 (0.85-1.11). Max L distal S 80 (61-90). IVo as in *P. flavum*. Microtrichia on dorsal surface usually more extensive proximally and along medial margin. L IVo 103 (86-115). W IVo 37 (34-45) (9). IVo dorsal S 25.5 (23-32), moderate to weakly developed. Gonostylus as in *P. flavum*, medial setae 9.4 (7-11), numerous, but weakly developed. Max L medial S 35 (32-38) (9). Max L apical S 30 (25-34) (8). L gonostylus 166 (146-189). GsR 3.32 (2.78-4.42). HR 2.91 (2.60-3.50). L posterior lobe gonocoxite 57 (52-65). HV 2.19 (1.84-2.38). Phallapodeme as in *P. flavum*.

109

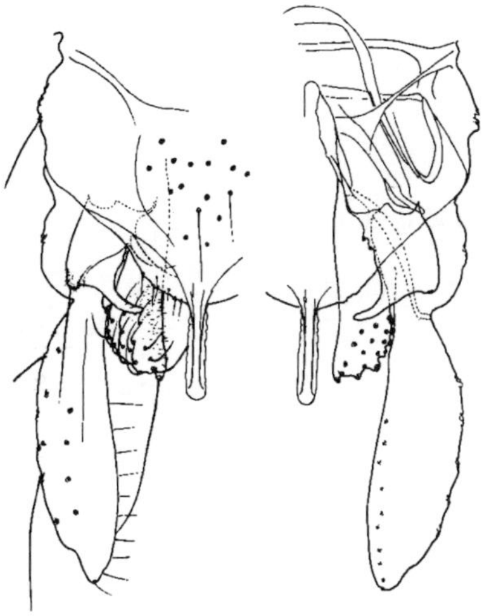

110

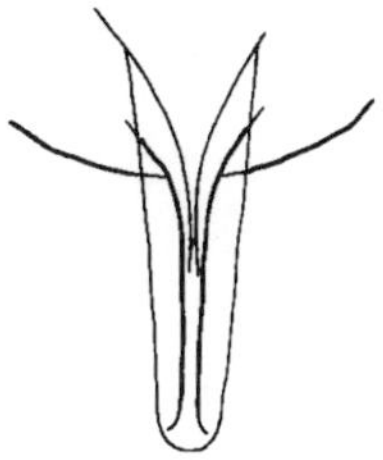

111

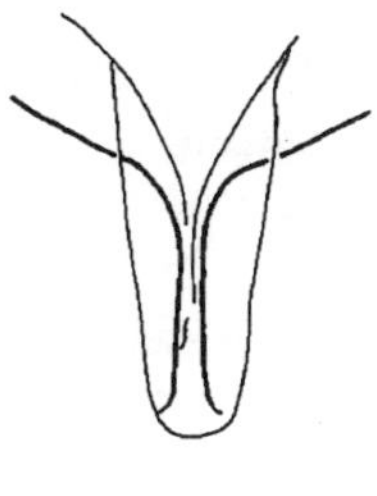

Figures 109-111. *P. obtusum* Townes: 109-hypopygium; 110 & 111-variations of anal point.

Type locality: **New York**, Hancock, 24 July 1935, H.K. Townes, at light, male (HKT).

Larva

Diagnosis—Indistinguishable from *P. flavum*, except possibly by Lauterborn organs exceeding apex of antennal segment 3.

Description—(n=6; 5 males, 1 female, female associated through rearing that included a male)

Color: Pale yellow-green (Sublette, 1957a).

Eye spots: Contiguous (Sublette, 1957a).

Biometrics: L total 5.4 mm (Sublette, 1957a) L head capsule 197 (171-214)

Mentum: W 102 (95-110), otherwise as in *P. flavum*.

Ventromental plates: W 82 (74-90), L 43 (38-47). VmPR 1.89 (1.78-2.02). Distance between ventromental plates 57 (52-61). VmPSR 1.44 (1.32-1.58). Striations as in *P. flavum*.

Antenna: (Figure 112). L segments 1-5: 75 (71-79), 28 (26-30), 7 (6-8), 8 (8-8), 5 (5-5). AR 0.66 (0.64-0.68) (4). L primary blade 49 (45-52) (4), apex in relation to apex of segment 5 variable. L secondary blade 11 (9-13) (4). L Lauterborn organ 6 (5-6) (4), very transparent, not visible on some specimens; appearing to exceed apex of segment 3. Triangular structures visible; PS as in *P. flavum*. ROR 6.43 (5.64-6.90) (3).

Labrum-epipharynx: As in *P. flavum*.

Mandible: L 120 (110-124). L seta subdentalis 18 (18-20) (5), otherwise as in *P. flavum*.

112

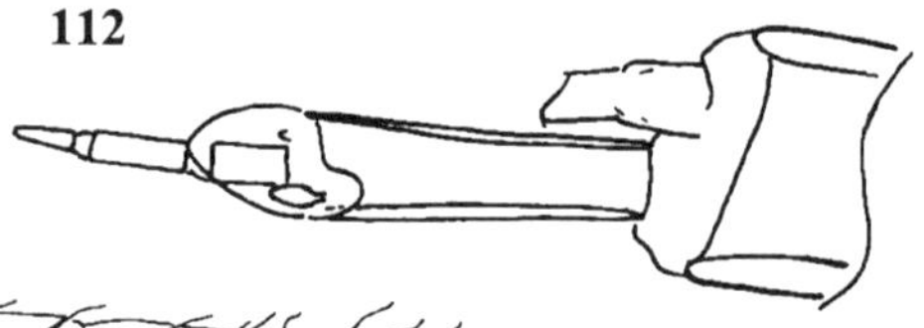

113

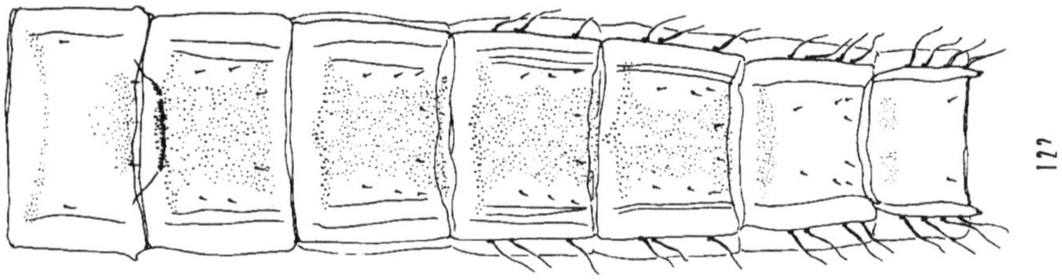

Figures 112-113. *P. obtusum* Townes: 112-antenna, segments 2-5, full length of primary blade not shown; 113-pupal abdomen.

Pupa

Diagnosis—Essentially same as *P. flavum*. Pedes spurii B present on segment II. Spinule pattern on tergites III to VI not separated laterally by bare midline.

Description—(n=6; 5 males, 1 female)

Biometrics: L total 4.06 (3.64-4.69) (5) mm L cephalothorax 1.12 (1.04-1.23) (5) mm

Coloration: Typical, with some darker specimens having light brown coloring anteriorly and laterally on segments II through VI.

Abdomen: (Figure 113). As in *P. flavum* except as noted in diagnosis. Anal spur with 1 to 3 apical accessory points. Anal taeniae 63.4 (43-74) on each anal lobe.

Remarks

There seem to be two rather distinct populations within this species with few intermediate specimens. One group has narrow, parallel-sided lateral expansions on the anal point; the other has broad, rounded lateral expansions on the anal point. Otherwise, individuals in the two groups are essentially identical, and we feel it is not justified to recognize them as separate species based on the anal point characters alone. The paratype examined has the broad anal point. Males of all the reared and mass-reared material, and most of the material examined collected in light traps in Minnesota, have the narrow anal point.

Boesel (1985) synonymized *P. obtusum* with *P. convictum* based on the variability of the anal point and the inability to distinguish the immature stages. Oliver *et al.* (1990) listed the synonymy. As specimens of the two species have not been reexamined, we are not recognizing the synonymy in this work.

Biology

Polypedilum obtusum is not commonly collected. We reared specimens from larvae collected on submerged logs in a trout stream (Sucker Brook), and adults were mass reared from algae and plant debris caught in overhanging branches at the water's surface (headwaters of Mississippi River). The larvae described by Sublette (1957a) were collected from moss covered gravel-rubble substrate in a small, clear stream. Beck (1977) summarized the environmental requirements of this species.

Material Examined

—Reared Specimens

Illinois, Hart Woods Creek, Mahomet, Champaign Co., 26 April 1967, D.W. Webb, 3 males, 1 female, (INH). **Louisiana**, Little Bayou Pierre, Kasatchie Wold, several dates in March and May 1956, J.E. Sublette, 10 males, 16 females (JES). **Minnesota**, Sucker Brook, Clearwater Co., 47°16'N 95°16'W, 3 Aug. 1969, D.E. Maschwitz, 1 male, 2 females, (UMn). **Oklahoma**, Briar Creek, Marshall Co., 25 June 1962, J.E. Sublette, 2 males, (JES). Records from April through June and August.

—Adult Specimens

Paratype male, **New York**, Ithaca (Corn). **Connecticut**, Norwich, 9 June 1959, A. Rubert, light trap, 1 male, (USNM). **Illinois**, Hart Mem. Woods, Mahomet, Champaign Co., 10-11 June 1966, M.W. Sanderson, 1 male, (INH). **Massachusetts**, Sunderland, 21 June 1951, at light, 1 male, (JES). **Minnesota**, Lower LaSalle Lake, Hubbard Co., from 17 June to 8 Sept. 1970, New Jersey mosquito light trap, 19 males, (UMn); New London Fish Hatchery, Kandiyohi Co., 28 June and 9 Aug. 1969, New Jersey mosquito light trap, 2 males, (UMn); Crookston Experiment Station, Polk Co., 14 July 1969, 16 July 1970, New Jersey mosquito light trap, 2 males, (UMn); U.S. EPA Lab., Duluth, St. Louis Co., 23 July and 12 Aug. 1970, New Jersey mosquito light trap, 2 males, (UMn); Blue Mounds State Park, Rock Co., 29 June 1970, New Jersey mosquito light trap, 1 male, 10 Aug. 1970, New Jersey mosquito light trap, 2 males, (UMn); Mississippi R. 6 mi. N Itasca State Park, Clearwater Co., 11 and 23 July 1970, New Jersey mosquito light trap, 2 males, (UMn); Mississippi R., near headwaters, Itasca State Park, Clearwater Co., 12 Oct. 1969, mass reared, D.E. Maschwitz, 2 males, (UMn); Minneapolis, 4 June 1959, D.E. Maschwitz, at light, 1 male, (UMn). **New York**, 1 mi. E Silver Creek, Lake Erie, Chautauqua Co., 19 Sept. 1967, D.W. Webb, 1 male (INH). **Ohio**, Little Miami R. Milford, reared, June, July, and Aug., W.T. Mason 10 males (EPACin); Little Miami R., Batavia, Clermont Co., reared, 16 Jan. 1969, W.T. Mason, 3 males (EPACin). **Virginia**, Vesuvius, Rockbridge Co., July 1960, D.H. Messersmith, light trap, 10 males (USNM); Vesuvius Nature Camp, 15 July 1960, D.M. Messersmith, 3 males, (USNM); Holmes Run, Falls Church, light trap, 20 June 1960 and 29 June 1961, W.W. Wirth, light trap, 2 males, (USNM). **Ontario**, Rideau River, Ottawa, 29 May 1960, W.W. Wirth, 1 male, (USNM). **Quebec**, Ile Ste Helene, Montreal, 24 July 1965, P.S. Corbet, 5 males (CNC). Records from June through September.

Polypedilum (P.) ontario (Walley) 1926

Chironomus ontario Walley, 1926: 206

Polypedilum ontario (Walley) Townes, 1945: 51; Beck, 1977 (biology); Pinder and Reiss, 1983: 335, 413 (figure of larva); Boesel, 1985: 258; Pinder and Reiss, 1986: 362, 453 (Chironomini Genus C, figures of pupa); Oliver *et al.* 1990: 51; Bolton, 1991: 125 (identification of Chironomini Genus C); Epler, 1995: 7.95.

Chironomus hirtipes Mitchell, 1908: 9 (*fide* Townes, 1945: 51, name preoccupied); Johannsen, 1908: 280 (female).

Adult Male

Diagnosis—*Polypedilum ontario* is perhaps the most distinctive Nearctic species in the genus. Thoracic and abdominal setae are numerous and long; abdominal setae concentrated in brown transverse bands on abdominal tergites II-V give a tufted appearance to the abdomen. Anal point broadened and darkly colored. Gonostylus small.

Description—(n=7)

Coloration: Head, pedicel, flagellum, femora, tibiae, forecoxae, most of thorax, and haltere light brown, tarsi stramineous. Maxillary palps usually shade darker than rest of head. Ventral two-thirds of preepisternum, anterior one-fourth of postnotum, mid and hind coxae, brown. Anterior two-thirds of abdominal tergite I, broad transverse band on abdominal tergites II-V, and all of tergites VI-VIII, brown. Gonostyli light brown, anal point dark brown, remainder of hypopygium brown to light brown. Wing uniformly light brown except for stramineous base.

Biometrics:	L total 5.69 (4.49-6.25) (6) mm	L thorax 1.68 (1.44-1.95) (6) mm
	L wing 2.86 (2.63-3.20) mm	W wing 0.83 (0.76-0.93) (6) mm

Antenna: AR 1.22 (1.17-1.29) (5). D scape 218 (200-240) (4). L pedicel 148 (140-170) (5). L flagellomeres 1-12 716 (670-790) (5). L flagellomere 13 874 (830-960) (5).

Head: W 770 (710-830) (4). Interocular distance 158 (150-170) (4). HIR 4.87 (4.73-5.07) (4). Vertical S 17.0 (1). Clypeal S 58 (1). Ocelli absent. Maxillary palp segment 3 slightly shorter than segment 4. L maxillary palp segments 1-5: 71 (65-79) (3), 77 (61-90) (3), 199 (171-223) (4), 213 (196-227) (4), 298 (275-320) (4).

Thorax: Dorsal edge of antepronotum as in *P. fallax*, usually about even with anterior margin of scutum. Antepronotal campaniform sensillum and chaeticum sensillum apparently absent. Postpronotal depression relatively small for *fallax* group species, sensilla not distinct. Pronotal S 7.2 (5-9) (6). Humeral pit absent. Acrostichal S 39 (1). Dorsocentral S 52.9 (41-75) (5). Some specimens with scattered anterior setae; main part of row irregularly double anteriorly to triple or quadruple posteriorly. Scutal fossal S 3.1 (1-6) (5). Prealar S 18.0 (12-21) (6). Scutellar S approximately 60, in 3 to 5 irregular rows, anterior setae smaller.

Wing: VR 1.19 (1.16-1.22). Apex of R_{4+5} equidistant to somewhat closer to wing apex than apex of M. R_1 and R_{2+3} slightly separated to adjacent apically. Brachiolar S 7.2 (6-9) (6). R_1 S approximately 100 (87-105) (3). R_{4+5} S 74 (63-84) (4). Squamal S approximately 33 (30-35) (2).

Legs: Femora and tibiae densely setate. Foretibial apical scale as in *P. fallax* but with apex often more rounded and not darkened. L foretibial apical scale 77 (70-86). L mesotibial apical spine 27 (23-34). L hindtibial apical spine 26 (18-34) (5), apex truncate.

	Fe	T1	Ta1	Ta2	Ta3
PI	1471 (1390-1690)	870 (770-1030)	1723 (1640-1870) (4)	1135 (1060-1240) (4)	845 (780-940) (4)
PII	1794 (1660-2080)	1330 (1230-1570)	760 (680-840) (6)	530 (470-580) (6)	388 (360-430) (6)
PIII	1879 (1760-2180)	1471 (1370-1700)	1200 (1140-1280) (3)	710 (670-770) (3)	607 (580-660) (3)

	Ta4	Ta5	LR	BV	SV
PI	668 (630-740) (4)	255 (230-280) (4)	1.92 (1.81-2.04) (4)	1.45 (1.41-1.48) (4)	1.41 (1.36-1.46) (4)
PII	282 (260-310) (6)	152 (140-170) (6)	0.56 (0.53-0.59) (6)	2.96 (2.76-3.09) (6)	4.19 (4.00-4.38) (6)
PIII	403 (380-440) (3)	183 (170-200) (3)	0.83 (0.82-0.84) (3)	2.38 (2.34-2.43) (3)	2.74 (2.71-2.77) (3)

Hypopygium: (Figure 114). Tergite IX S 27.2 (21-42) (6), located in large middorsal oval area. Anal point with broad base, usually becoming slightly broader midway along its length then narrowing apically.

IVo exceeding anal point posteriorly, -46 (-18 to -86). SVo narrow and evenly curved medially or fairly broad and curved medially posterior to origin of distal seta. Apical portion of SVo narrowed, without terminal knob. Medial margin and most of dorsal surface of basal portion bearing microtrichia, Basal S 6.2 (4-9) (6). LSVo 149 (117-167). Placement of distal seta somewhat variable, mean SVoR 0.72 (0.45-1.10). Max L distal S 137 (70-176) (5), usually exceeding IVo posteriorly. Small distal seta appears to be associated with SVoR near 1.0. IVo long and narrow, with large subapical dorsal swelling bearing over half of dorsal setae. Apex usually with 2 or 3 setae directed posteriorly but curving dorsally. IVo dorsal S approximately 27 (25-29) (2), moderately to poorly developed. L IVo 228 (197-266) (6). W IVo (not including dorsal swelling if turned laterally) 32 (27-38) (3). Gonostylus small, narrowed apically. Medial S 6.0 (5-7), concentrated on posterior 1/3 of gonostylus. Max L medial S 123 (108-146) (5), usually reaching the midline. Max L apical S 52 (45-61) (3). L gonostylus 240 (218-268). GsR 3.76 (3.13-4.91). HR 1.83 (1.67-1.98). L posterior lobe gonocoxite 131 (124-142). HV 2.36 (2.26-2.62). Basal plate and medial ridge of gonocoxite distinct. Phallapodeme broadened midway along its length.

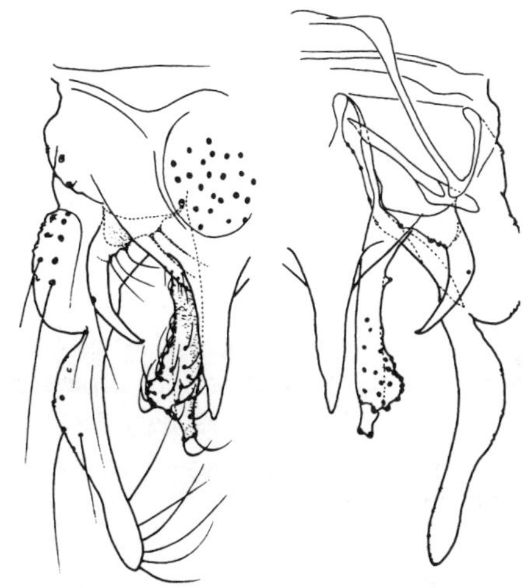

Figures 114. *P. ontario* (Walley): 114-hypopygium.

Type locality: *P. ontario*, **Ontario**, Ottawa, 11 Aug. 1924, G.S. Walley, male, (CNC); *P. hirtipes*, **Maryland**, Cabin John, 3 June, male, (USNM).

Larva

Diagnosis—Head capsule narrowed anteriorly; mouth parts unusually small; striations on ventromental plates indistinct. (Figures 115, 116)

115

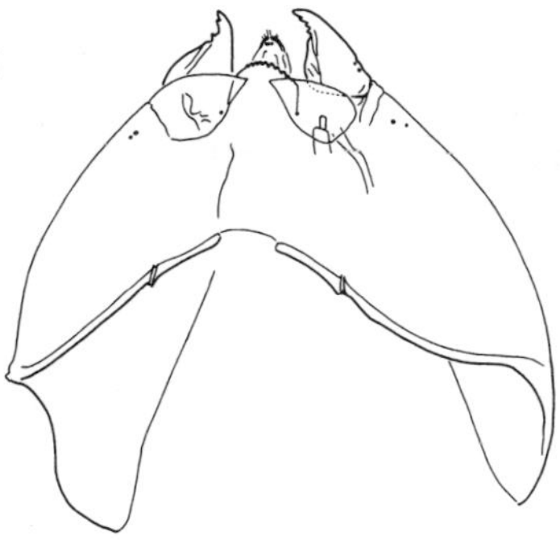

116

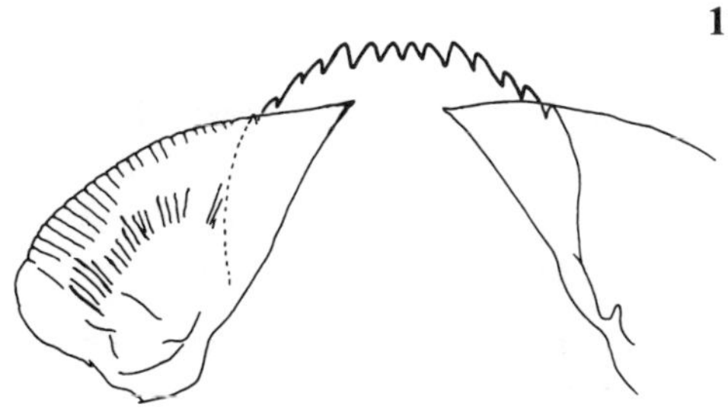

Figures 115-116. *P. ontario* (Walley): 115-head capsule, ventral view; 116-mentum.

Pupa

Diagnosis—Cephalic tubercles cone shaped and darkened at tip. Abdominal tergites without anterior row of large spinules. (Figures 117-119).

Description—See Sæther and Sundal (1999), Pinder and Reiss (1986: 362 - description of Chironomini Genus C), and Bolton (1991).

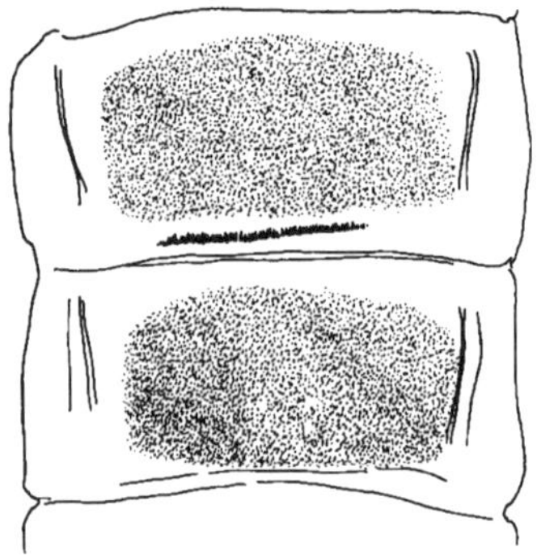

117

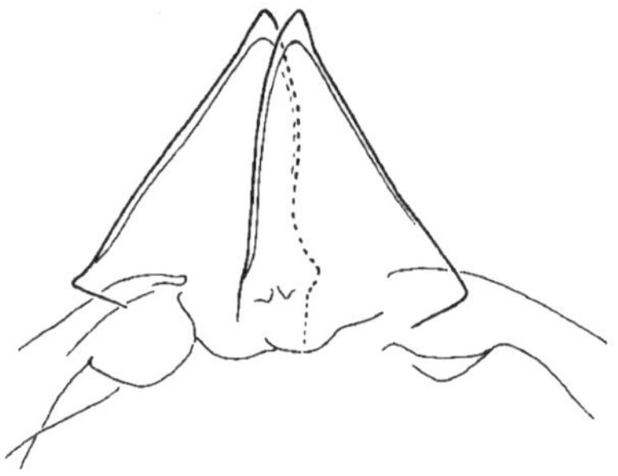

118

119

Figures 117-119. *P. ontario* (Walley): 117-pupal abdomen, tergites II and III; 118-cephalic tubercles; 119-anal spur on tergite VIII.

Remarks

This species has characteristics atypical of the genus in the adult male, larva, and pupa. The uniqueness of *P. ontario* has led some to suggest that it may not belong in the genus *Polypedilum* (Pinder and Reiss, 1983; Pinder and Reiss, 1986; Epler, 1995). Saether and Sundal (1999) have designated a new subgenus in *Polypedilum*, *P.* (*Cerobregma*), to include *P. ontario* and related species world wide. *Polypedilum ontario* is the only known Nearctic representative in the new subgenus. Subgeneric status for the *P. ontario* group of species seems justified due to their unique characters. However, Saether and Sundal (1999) appeared too late to include the new subgenus in this work (see Authors' Note, page v). *Polypedilum ontario* is not commonly collected.

Biology

The larvae of this species live in the pupal cases of the caddisfly *Cheumatopsyche* (Bolton, 1991).

Material Examined—Adult Specimens

Arkansas, White River, Hwy 68 bridge, Washington Co., SE 1/4 T17N R29W, 13 and 25 July 1963, L.R. Aggus, light trap, 2 males, (JES). **Maryland**, Jackson's Island, 22 June 1902, H.S. Barber, 2 males, (USNM). **Minnesota**, Lower LaSalle Lake, Hubbard Co., 2 Sept. 1970, leg. E.F. Cook, New Jersey mosquito light trap, 1 male, (UMn). **New York**, Oneonta, 5 July 1935, H.K. Townes, 1 male, (Corn). **Pennsylvania**, Hazleton, 24 June 1909, D.R. Dietz, 1 male, (Corn). Records from June through September.

Polypedilum (*P.*) *ophioides* Townes 1945

Polypedilum ophioides Townes, 1945: 57; Berg, 1950: 92 (misidentification); Roback, 1957: 115 (key to larva, based on Berg's description of *P. ophioides* which is *P. bergi* sp. nov.); Sublette, 1960: 207; Harper and Cloutier, 1979 (biology); Boesel, 1985: 260.

Adult Male

Diagnosis—SVo very large, curved medially at point of distal seta origin. Distal portion of SVo perpendicular to approximately mid part of anal point. SVo basal setae in 2 groups; first group on base proper, second group on medial margin posterior to first.

Description—(n=6)

Coloration: Generally very light. Thorax, abdomen, head, haltere, wing veins stramineous. Legs stramineous except for light brown tinging on foretibia. Pedicel and flagellum light brown.

Biometrics:	L total 3.98 (3.66-4.34) mm	L thorax 1.14 (1.07-1.24) mm
	L wing 2.15 (1.96-2.35) mm	L wing 0.64 (0.61-0.68) mm

Antenna: AR 1.83 (1.75-1.88). D scape 168 (160-170) (4). L pedicel 113 (110-120) (3). L flagellomeres 1-12 458 (440-480). L flagellomere 13 835 (800-900).

Head: W 550 (530-590). Interocular distance 102 (100-110). HIR 5.39 (5.00-5.90). Vertical S 13.0 (11-14). Clypeal S 19.2 (14-22). Ocelli small, not visible on some specimens. L maxillary palp segments 1-5 44 (41-45) (5), 43 (34-50) (5), 129 (124-137) (5), 116 (110-122) (5), 200 (187-212).

Thorax: Antepronotum as in *P. illinoense*; thoracic sensilla similar to *P. illinoense*. Postpronotal depression with visible campaniform sensilla. Humeral pit apparently absent. Scutal fossal S usually present, 1.8 (0-4). Dorsocentral S 15.6 (12-19), uniserial. Prealar S 6.0 (5-7). Scutellar S in 2 rows, anterior row smaller; anterior row 8.0 (6-10); posterior row 11.5 (10-13).

Wing: VR 1.18 (1.15-1.19), wing unspotted. Apices of R_{4+5} and M about equidistant from wing apex. R_1 and R_{2+3} adjacent or slightly separated apically. Brachiolar S 1. R_1 S 42.0 (37-46). R_{4+5} S 36.3 (32-40). Squamal S 12.3 (11-14).

Legs: L foretibial apical scale 37 (34-38). L spine on foretibial apical scale 12 (11-14), usually tinged with light brown. L mesotibial apical spine 34 (34-34) (5). L hindtibial apical spine 38 (34-41).

	Fe	**T1**	**Ta1**	**Ta2**	**Ta3**
PI	962 (860-1000)	647 (590-700)	1137 (1050-1180)	855 (790-920)	578 (560-600)
PII	1072 (980-1150)	890 (830-960)	535 (510-560)	342 (320-380)	247 (230-260)
PIII	1138 (1050-1220)	1048 (980-1130)	780 (730-860) (5)	450 (420-490) (5)	368 (340-390) (5)

	Ta4	**Ta5**	**LR**	**BV**	**SV**
PI	430 (400-450)	180 (160-190)	1.78 (1.67-1.83)	1.35 (1.31-1.38)	1.40 (1.37-1.45)
PII	153 (140-160)	85 (70-90)	0.62 (0.59-0.68)	3.03 (2.97-3.17)	3.68 (3.55-3.81)
PIII	228 (220-240) (5)	108 (100-120) (5)	0.71 (0.66-0.75) (5)	2.62 (2.57-2.69) (5)	2.87 (2.76-2.96) (5)

Hypopygium: (n=7, including holotype) (Figure 120). Tergite IX S 5.1 (2-7), located in long narrow middorsal area. Tergite IX concave laterally at base of anal point. Anal point straight, slightly curved ventrally, not expanded laterally. Anal point not extending beyond apex of IVo, -22 (-11 to -38). SVo large, portion proximal to distal setae origin larger and extending further posteriorly than in any other species of *Polypedilum*. SVo curved medially at point of distal seta origin. Distal portion of SVo perpendicular to approximately mid part of anal point. SVo usually terminating in rounded knob. Basal setae usually separated into 2 groups: first on proximal part of base, second placed distally on medial margin. Setae in distal group, if more than 1, occasionally widely separated. Basal S in proximal group 2.4 (2-3) (7), in distal group 1.6 (1-2) (7). SVo medial margin and dorsal surface proximal of distal seta origin covered with microtrichia. LSVo 130 (119-140). Mean SVoR 0.89 (0.77-1.07). L distal S 107 (86-128) (6), usually exceeding apex of IVo. IVo similar to *P. illinoense*, L IVo 142 (133-158). W IVo 47 (38-54). Dorsal S 15.9 (13-18), most setae located near posterolateral margin. L gonostylus 196 (183-209), long and narrow. GsR 5.45 (4.47-6.00). Medial S 3.3 (2-5) (6), weak. Max L medial S 45 (38-54) (6). Max L apical S 24 (20-27) (6). L posterior lobe gonocoxite 56 (52-59). HR 3.53 (3.27-3.60). HV 2.06 (1.96-2.18) (6). Phallapodeme usually curved midway along its length, medial end usually slightly to moderately broadened.

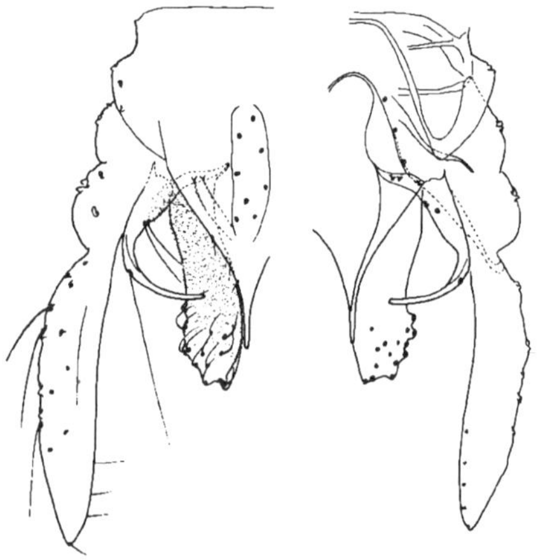

Figures 120. *P. ophioides* Townes, hypopygium.

Type locality: **New York**, banks of Cadosia Creek, Cadosia, 6 Aug. 1935, H.K. Townes, male, (HKT).

Larva

Diagnosis—Very close to *P. illinoense*; Lauterborn organs reaching apex of antennal segment 3.

Description—(n=3; all males)

Biometrics: L head capsule 200 (191-207)

Mentum: W 116 (106-126), otherwise as in *P. illinoense*.

Ventromental plates: W 101 (95-104), L 44 (43-45). VmPR 2.31 (2.24-2.40). Striations (approx.) 40 (1). Distance between ventromental plates 56 (52-59). VmPSR 1.81 (1.75-1.85).

Antenna: (Figure 121). L segments 1-5: 65 (64-67) (2), 21 (20-22) (2), 7 (6-7) (2), 9 (1), 4 (4-4) (2). AR 0.65 (1). L primary blade 49 (1), apex even with apex of segment 5. L secondary blade 10 (1). L Lauterborn organ 9 (8-9) (2), apex reaching apex of segment 3. Triangular structures visible at base of Lauterborn organs. ROR 4.97 (4.53-5.41) (2). Segment 1 with 2 sense pits, proximal sense pit indistinct but visible on some specimens.

Labrum-epipharynx: As in *P. illinoense*. PE with 4-5 teeth in lateral parts, 3 teeth in middle part.

Mandible: L 140 (135-144). L seta subdentalis 18 (16-20), otherwise as in *P. illinoense*.

Maxilla: L lacinia 39 (34-43). L segment 1 maxillary palp 18 (16-20).

Pupa

Diagnosis—Similar to *P. illinoense*, can be distinguished from this species by lack of pedes spurii A on tergite IV, lack of fenestrations in shagreen spinules, and larger spinules in anterolateral patches on tergite VII.

Description—(n=3; all males)

Biometrics: L total 4.41 (4.08-4.58) mm L cephalothorax 1.12 (1.00-1.20) mm

Coloration: Typical.

Abdomen: (Figure 122). Pedes spurii B on segments I and II reduced or absent. Pedes spurii A absent. Anterior rows of spinules on tergites II-VI larger than hooks on tergite II and intersegmental spinules. Intersegmental spinules between segments III and IV irregularly uniserial; those between segments IV and V irregularly biserial to multiserial. Intersegmental spinules pointing anteriorly. Shagreen spinules in rough triangular pattern on tergite II and roughly V shaped centrally on tergites III-VI. Also, tergites III-VI with broad posterior band of shagreen spinules. Shagreen spinules without fenestrations. Anterolateral patches of spinules on tergite VII with relatively large spinules; same spinules on tergite VIII small. Lateral taeniae 3 on segments V and VI; 4 on segments VII and VIII. Anal spurs with 1 to 2 large apical points and 1 to 2 smaller points. Anal taeniae 38.2 (35-43).

121

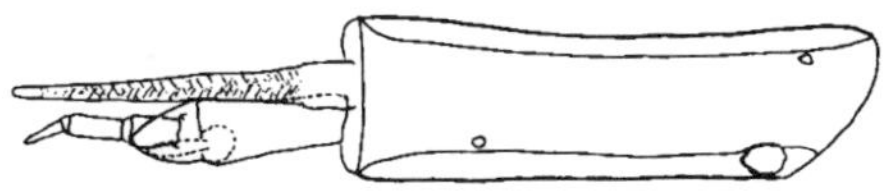

122

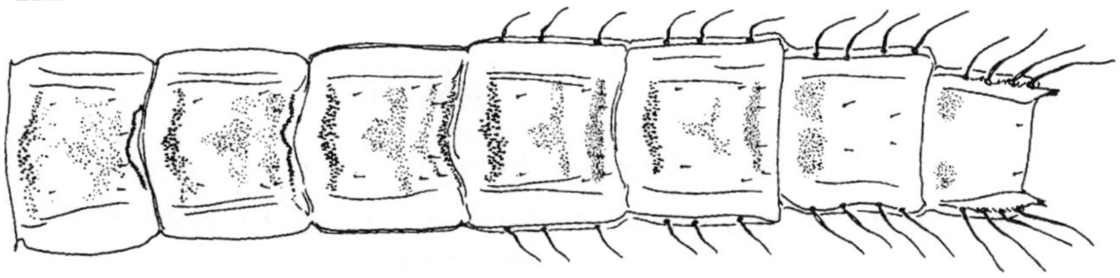

Figures 121-122. *P. ophioides* Townes: 121-antenna; 122-pupal abdomen.

Remarks

Epler (1995) states that the antennal characters used to separate larval *P. ophioides* from *P. illinoense* have not held up upon the examination of additional specimens.

Biology:

Harper and Cloutier (1979) report *P. ophioides* from the L'Achigan River, Quebec and from a fast flowing tributary, but in very small numbers.

Material Examined

—Reared Specimens

FWI Pipeline Project, Martin River, Northwest Territories, 4 June 1973, 3 males, (CNC).

—Adult Specimens

Holotype male, (slide of hypopygium). Paratype male, no data (Corn). **New York**, Lake Placid, July 1922, paratype male, J.M. Aldrich, (USNM). **Massachusetts**, Sunderland, 21 June 1951, at light, 1 male, (JES). **Minnesota**, New Jersey mosquito light trap material, leg. E.F. Cook: Minnesota Forestry Station, Little Fork, Koochiching Co., June, July, and August, 1968-1970, 37 males, (UMn); 1 mi. S Caledonia, Houston Co., 22 July 1968, 2 males, (UMn); Crookston experiment station, Polk Co., 14 July 1969, 24 June 1970, 2 males, (UMn); U.S.EPA Lab., Duluth, St. Louis Co., 20 Sept. 1969, 1 male, (UMn); Mississippi River at Co. Rd. 2, Clearwater Co. 13 June 1970, 2 males and 1 July 1970, 1 male, (UMn); Stillwater, Washington Co., 9 July, 1969, D.E. Maschwitz, at light, 1 male, (UMn). **New York**, Ithaca, April, O.A. Johannsen, 1 male, (Corn). **Pennsylvania**, Conewago Creek, 13 mi. W York, 20 June 1959, A. Hubert, light, 1 male; 19-21 June 1959, J.P. Spangler, at light, 1 male, (USNM). **Virginia**, Blacksburg, April 1960, D.H. Messersmith, light trap, 1 male, (USNM). **Wisconsin**, 11 mi E, 4 mi. S Siren, Burnett Co., 15 July 1966, 1 male; 16 July 1966, 3 males; 6 Aug. 1966, 4 males; D. Hansen, light trap, (UMn). **Montana**, Skalkano Divide Rocky mountain Lab, 22 July 1967, C.B. Philip, 1 male, (JES). Records from April and June through September.

Polypedilum (*U.*) *pedatum* Townes 1945

Polypedilum pedatum Townes, 1945: 55; Oyewo and Sæther, 1998.

Polypedilum pedatum excelsius Townes, 1945: 55; Townes, 1952: 52; Sublette, 1960: 207. New synonymy. Grodhaus and Rotramel, 1980 (biology, descriptions and figures of immature stages); Harper and Cloutier, 1979 (biology); Boesel, 1985: 259; Niitsuma, 1992 (description, figure of hypopygium); Sasa and Kikuchi, 1995.

Adult Male

Diagnosis—Dark species. AR 1.04. Distinguished from *P. sulaceps* by much larger gonostylar medial setae, and light colored scutellum.

Description—(n=6)

Coloration: Head, antenna, and thorax, except scutellum brown, scutellum and legs stramineous to light brown, haltere stramineous with tinge of brown. Abdominal tergites and hypopygium brown but shade lighter than thorax.

Biometrics:	L total 4.10 (3.40-5.04) mm	L thorax 1.12 (0.94-1.32) mm
	L wing 2.21 (1.75-2.72) mm	W wing 0.64 (0.52-0.76) mm

Antenna: AR 1.04 (0.83-1.19). D scape 153 (140-160) (4). L pedicel 117 (110-120) (3). L flagellomeres 1-12 555 (480-620). L flagellomere 13 572 (500-660).

Head: W head 490 (460-530) (5). Interocular distance 90 (80-100) (5). HIR 5.47 (5.00-6.00) (5). Vertical S 13.5 (12-15) (4). Clypeal S 16.5 (11-20). Ocelli very small, not seen on many specimens. L maxillary palp 1-5: 41 (34-45) (3), 37 (34-41) (4), 183 (135-276) (5), 151 (126-167) (5), 222 (182-261) (5).

Thorax: Prothorax as in *P. illinoense*, dorsal prothoracic sensillum distinct. Postpronotal depression small, campaniform sensilla not visible on most specimens. Humeral pit present. Dorsocentral S 17.8 (13-22) (5), uniserial in most specimens, but row becomes double posteriorly in large specimens. Prealar S 7.0 (5-11). Scutellar S in 2 rows, setae in anterior row small to nearly same size as setae in posterior row; anterior row S 5.3 (3-7); posterior row S 11.7 (10-15).

Wing: VR. 1.16 (1.10-1.19). Extent of downward curvature of R_{4+5} variable. Apices of R_{4+5} and M nearly equidistant from wing apex when R_{4+5} curvature is slight (Townes' subspecies *pedatum*). Apex of R_{4+5} at wing apex when curvature is pronounced (Townes' subspecies *excelsius*). R_1 and R_{2+3} adjacent or separated by a small gap apically. Brachiolar S 1.4 (1-2) (5). R_1 S 43.2 (29-64) (5). R_{4+5} S 51.8 (24-58). Squamal S 11.6 (9-18) (5).

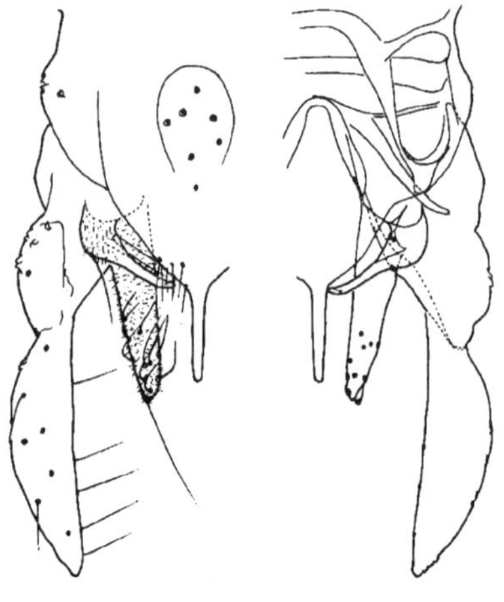

Figure 123. *P. pedatum*, hypopygium.

Legs: Foretibial apical scale without spine. L apical scale 38 (32-43). L mesotibial apical spine 18 (14-25). L hindtibial apical spine 18 (14-23) (5).

	Fe	T1	Ta1	Ta2	Ta3
PI	987 (810-1190)	843 (660-1100)	1013 (850-1140)	648 (560-740)	527 (420-630)
PII	1103 (920-1350)	962 (800-1180)	530 (430-650)	318 (250-410)	247 (190-320)
PIII	1183 (980-1450)	1098 (900-1340)	787 (660-960)	445 (350-580)	397 (310-520)
	Ta4	**Ta5**	**LR**	**BV**	**SV**
PI	390 (310-470)	170 (130-210)	1.22 (1.04-1.29)	1.65 (1.60-1.69)	1.80 (1.73-2.01)
PII	157 (120-190)	102 (80-130)	0.55 (0.53-0.57)	3.18 (3.05-3.36)	3.92 (3.81-4.02)
PIII	237 (200-290)	118 (90-150)	0.71 (0.70-0.73)	2.59 (2.45-2.67)	2.92 (2.85-2.99)

Hypopygium: (Figure 123). Tergite IX S 10.2 (7-16) (5), in middorsal egg-shaped area. Anal point straight, not expanded laterally. Anal point not exceeding apex of IVo, -11 (-5 to -23). SVo with broad basal portion, apical portion angled medially, apex usually curved anteriorly. Microtrichia on ventral and dorsal surface of basal portion of SVo. Microtrichia on medial margin numerous and large. Basal S 2.1 (1-3) (5). LSVo 97 (86-122). Mean SVoR 1.54 (1.16-2.15). Max L distal S 61 (50-79) (5). IVo usually tapered, broadest at base, apex rounded or bluntly pointed, medial and lateral margins usually irregular. Apical or subapical posteriorly directed seta present. Microtrichia on IVo dorsal surface distinct. L IVo 110 (104-115) (5). W IVo 30 (23-43) (5). IVo dorsal S 13.3 (10-17). Gonostylus, medial margin straight to slightly convex. Medial S 4.5 (3-6), moderately well developed, usually not reaching midline. Max L medial S 64 (50-86). Max L apical S 37 (34-52). L gonostylus 142 (124-171). GsR 3.21 (2.61-3.61). HR 1.86 (1.63-2.00). L posterior lobe gonocoxite 76 (70-86). HV 2.94 (2.81-3.15). Phallapodeme expanded toward medial end.

Type locality: *P. pedatum pedatum*, holotype male, **New York**, McLean Bogs Reserve, McLean, (Corn). *P. pedatum excelsius*, holotype male, **Washington**, Lake Cushman, 27 June 1917, H.G. Dyar, (USNM).

Larva and Pupa

Description—See Grodhaus and Rotramel (1980)

Remarks

Townes (1945) recognized two subspecies in this species, the eastern *pedatum* and the western *excelsius*, based on the extent of downward curvature of the apical part of wing vein R_{4+5}. However, the intermediate condition is fairly common and some western specimens have only a slight curvature and some eastern specimens show an extensive curvature. The two Minnesota specimens have a well developed downward curvature of R_{4+5}. Therefore, we synonymized the subspecies *P. pedatum excelsius*.

Biology

Grodhaus and Rotramel (1980) discovered the immature stages of *P. pedatum* in dry litter collected from tree-holes (in laurel and oaks) in Marin and Sonoma Counties, California. Eggs, and at least the earlier instar larvae, are able to survive a certain amount of drying of the tree-hole litter. Larvae return to activity when wetted. The blood-red larvae showed a strong tendency to burrow into the substrate in the laboratory. Harper and Cloutier (1979) report moderate numbers of *P. pedatum* from a few L'Achigan River stations in Quebec.

Material Examined—Adult Specimens

P. pedatum excelsius, holotype male, (slide of hypopygium), and paratype male, **Washington**, Bremerton, S.S. Cook, (USNM). *P. pedatum pedatum*, holotype male (slide of hypopygium) and 2 paratype males, New York, McLean Bogs Reserve, McLean (Corn). **Virginia**, Mountain Lake, pr. #1, 11 July 1940, L.J. and M.J. Milne, paratype male, (USNM). **California**, Leavitt Meadow, Mono Co., Alt. 7200 ft., 13 Aug. 1963, H.B. Leach, flight trap, 3 males (CalAS). **Minnesota**, 2 mi. N Hovland on U.S. 61, Cook Co., 2 June 1970 and 16 Sept. 1969, E.F. Cook, New Jersey mosquito light trap, 2 males, (UMn). **New York**, Cow Neck, Long Island, spring seep, 24 May 1963, W.W. Wirth, 6 males, (USNM); Singing Waters Picnic Area, Lewis Co., 21 June 1963, W.W. Wirth, 1 male, (USNM); Fish Creek Pond, Franklin Co., 26 June 1963, W.W. Wirth, light trap, 2 males, (USNM); Sinclairville, Chautauqua Co., muddy brook, 31 May 1963, W.W. Wirth, 1 male, (USNM). **Oregon**, Eel Creek For. Camp, Coos Co., 31 Oct. 1961, light trap, E.L. Bessel, 1 male, (USNM). **Virginia**, Holmes Run, Falls Church, 17 Oct. 1960, W.W. Wirth, light trap, 1 male, (USNM). **Washington**, shore of Twin Lakes, Mt. Baker Nat. For. 27 Aug. 1966, D. Hansen, 2 males, (UMn); 12 mi. E, 3 mi. S Glacier, Artists' Pt., Mt. Baker, 12 Sept. 1967, D. Hansen, 1 male, (UMn). **British Columbia**, NW Bay, Marion Lake, 7 May 1965, A.L. Hamilton, 3 males, (FRB); Marion Lake, West Sabre, 25 May 1965, A.L. Hamilton, 2 males, (CNC). **Quebec**, Meach Lake, 6 June 1960, W.W. Wirth, 1 male, (USNM). Records from May through October.

Polypedilum (*P.*) *pedestre* (Meigen) 1830

Chironomus pedestris Meigen, 1830: 246; Kertesz, 1902: 197 (reference to early literature); Johannsen, 1905: 219; Goethebuer, 1921: 135 (description and figure of hypopygium).
Polypedilum pedestre (Meigen) Kieffer, 1921c: 73 (generic position); Goetghebuer, 1928: 91: Kruseman, 1933:133 (? midpart of fore femur dark); Goetghebuer, 1934: 393; Goetghebuer, 1936: 75 (biology); Goetghebuer, 1937: 62; Johannsen, 1937: 31 (generic position); Brundin, 1949: 226, 770; Thienemann, 1954: several pages; Dittmar, 1955: 482 (biology); Reiss, 1968: 266 (biology); Lehmann, 1971: 501 (biology); Moller Pillot, 1978-1979: IV.II.16; Pinder and Reiss, 1983: 411 (figures of larva); Rossaro, 1985: 20 (figures of larva and pupa); Pinder and Reiss, 1986: 425 (figures of pupa); Ashe and Cranston, 1991 (Palaearctic species); Oliver and Dillon, 1994; Sasa and Kikuchi, 1995: 240 (figure of hypopygium).
Chironomus fuscipennis Meigen, 1818: 35 (*nomen dubium*); Walker, 1856: 164; Goetghebuer, 1923: 127 (Meigen's type unidentifiable); Ashe and Cranston, 1991: 306 (listed as a questionable synonym of *P. pedestre*).
Chironomus (*Microtendipes*) *fuscipennis* Meigen; Edwards, 1929: 400; Sasa and Kikuchi, 1995: 34, 37, 38 (listed as a synonym of *P. pedestre*).
Polypedilum (*Polypedilum*) *fuscipenne* Meigen, Townes, 1945: 54 (*nomen dubium*)
Chironomus pictipes Zetterstedt, 1850: 3569 (*fide* Edwards, 1929: 400).
Chironomus patens Walker, 1856: 170 (*fide* Goetghebuer, 1928: 91).

Adult Male

Diagnosis—Very close to *P. fallax* but generally larger. Distinguished from *P. fallax* by following characters: apical one-half of foretibia and haltere brown to light brown, AR close to 1.0. HR approximately 2.0. IVo long.

Description—(n=4)

Coloration: Head, pedicel, and thorax brown to dark brown, scutellum and maxillary palps light brown, haltere brown to light brown. Flagellum and legs stramineous, except for brown to light brown color on parts of fore and hind coxae and distal one-half of forefemur. Wing veins stramineous. Abdominal tergites and hypopygium as in *P. fallax*.

Biometrics:	L total 5.65 (5.33-5.98) mm	L thorax 1.57 (1.51-1.62) mm
	L wing 2.66 (2.15-3.04) mm	W wing 0.81 (0.66-0.90) mm

Antenna: AR 1.07 (1.05-1.08). D scape 208 (190-220). L pedicel 140 (140-140) (2). L flagellomeres 1-12 685 (650-720) (2). L flagellomere 13 730 (680-780) (2).

Head: W 648 (640-660). Interocular distance 90 (80-100). HIR 7.20 (6.50-8.00). Vertical S 21.0 (20-22). Clypeal S 35.3 (32-40). Ocelli very small or absent. L maxillary palp segments 1-5: 62 (61-63), 67 (63-74(, 200 (1), 184 (178-189) (2), 297 (263-331) (2).

Thorax: Dorsal edge of antepronotum only slightly produced anteriorly, more like *P. illinoense* than *P. fallax*; otherwise antepronotum and thoracic sensilla as in P. fallax. Acrostichal S 22 (1). Scutal fossal S 3.8 (3-4) (3), dorsocentral S 22.0 (20-24) (3), uniserial, double at extreme posterior end of row. Prealar S 8.0 (6-9) (3). Scutellar S 31.3 (28-35), in 2 to 3 regular to irregular rows, anterior row of setae smaller.

Wing: VR 1.23 (1.19-1.29). Wing unspotted. Apex of R_{4+5} closer to wing apex than apex of M. R_1 and R_{2+3} adjacent apically. Brachiolar S 1. R_1 S 47.3 (34-55). Squamal S 20.8 (11-34).

Legs: Foetibial apical scale as in *P. fallax*. L foretibial scale 76 (70-79) (3). L mesotibial apical spine 27 (23-32). L hindtibial apical spine 32 (29-34).

	Fe	T1	Ta1	Ta2	Ta3
PI	1273 (1230-1320)	1003 (970-1030)	1400 (1)	950 (1)	720 (1)
PII	1505 (1430-1570)	1243 (1190-1290)	717 (700-750) (3)	437 (420-450) (3)	343 (340-350) (3)
PIII	1683 (1640-1740)	1457 (1420-1490) (3)	1040 (1030-1050) (2)	580 (1)	480 (1)

	Ta4	Ta5	LR	BV	SV
PI	600 (1)	230 (1)	1.45 (1)	1.45 (1)	1.57 (1)
PII	227 (220-240) (3)	130 (130-130) (3)	0.57 (0.56-0.58) (3)	3.04 (3.00-3.10) (3)	3.83 (3.83-3.84) (3)
PIII	300 (1)	150 (1)	0.73 (1)	2.73 (1)	2.98 (1)

Hypopygium: (Figure 124). Tergite IX S 12.3 (10-13), located in middorsal oval-shaped area. Anal point as in *P. fallax*, terminating short of IVo apices, -28 (-20 to -38). SVo as in P. fallax. Basal S 2.9 (2-3). LSVo 130 (119-140). Mean SVoR 0.78 (0.62-1.14). Max L distal S 171 (158-191), exceeding IVo posteriorly. IVo as in *P. fallax* only longer, apical or subapical posteriorly directed seta present. L IVo 202 (189-214). W IVo 37 (32-45). IVo dorsal S 23.6 (20-29), moderately well developed. Gonostylus as in *P. fallax*. Medial S 4.5 (4-5), fairly evenly spaced and in a straight row. Max L medial S 95 (88-101), generally ending near midline. Max L apical S 52 (1). L gonostylus 224 (214-232). GsR 4.00 (3.80-4.29). HR 2.02 (1.92-2.10). L posterior gonocoxite 110 (106-115). HV 2.59 (2.45-2.80). Phallapodeme as in *P. fallax*.

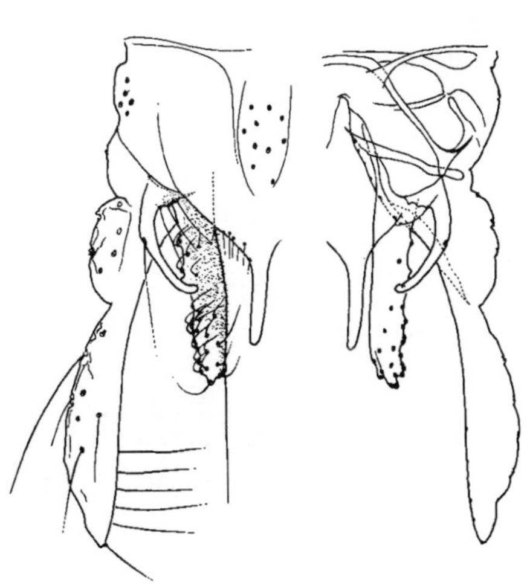

Figure 124. *P. pedestre* (=*P. fuscipenne* (Meigen?), hypopygium.

Type locality: *P. pedestris*, locality unknown, male, (? Paris Museum).

Remarks

Edwards (1929) indicated that *P. pedestris* (Meigen, 1830) is a synonym of *C. fuscipennis* (Meigen 1818). Unfortunately, *C. fuscipennis* was described from a female (Paris Musuem), and, according to Goetghebuer (1923), the type is unidentifiable. Townes (1945) compared his Nearctic material of *P. fuscipenne* to a male specimen from England determined by Edwards. Townes did not specifically state what this seems to imply, which is *P. pedestre* and *P. fuscipenne* are the same species, and the fact that he did not list *P. pedestre* as a synonym of *P. fuscipenne* indicates otherwise. More recent authors consider *P. pedestre* as a valid Holarctic species (*e.g.* Pinder and Reiss, 1983; Rossaro, 1985; Pinder and Reiss, 1986; Oliver and Dillon, 1994). Ashe and Cranston (1991) list *P. fuscipenne* as a questionable synonym, and Sasa and Kikuchi (1995) list *P. fuscipenne* as a synonym of *P. pedestre*. The figure of the *P. pedestre* hypopygium in Sasa and Kikuchi (1995) could as well be *P. fuscipenne*, but this does not resolve the question because the hypopygia of *fallax* group species tend to be very similar. If they are indeed the same species, and it may be impossible to determine due to the age, sex, and condition of the types, *P. fuscipenne* is the senior synonym and would be the valid name. However, until Holarctic material is examined more thoroughly, it seems best to consider *P. fuscipenne* (*sensu* Townes, 1945) a *nomen dubium* and use the name *P. pedestre* (Meigen). *Polypedilum fuscipenne* Kieffer (1921b) is a homonym of *P. fuscipenne* (Meigen).

Biology

Polypedilum pedestre is a common inhabitant of rivers and streams (Goetghebuer, 1936; Lehmann, 1971), and it apparently lives in the bottom sediments (Thienemann, 1954; Dittmar, 1955). Reiss (1968) found this species in sandy mud in Lake Constance, but states that it might have been carried in by incoming rivers. Harnisch (1922) has recorded this species swarming.

Material Examined—Adult Specimens

Washington, Dayton to Walla Walla, June 1950, R.B. Spurrier, at dusk, 1 male, (WashSU). **Wyoming**, S Fork Clear Creek, south fork campground, 16 mi. WSW Buffalo, 27 Aug. 1967, D. Hansen, sweeping by stream at dusk, 1 male (UMn). **British Columbia**, Hixon, 19-27 July 1966, 3 males (CNC). Records from June to August.

Polypedilum (P.) prolixipartum New Species

Adult Male

Diagnosis —Similar to *P. vibex* in coloration of abdominal tergites and very long IVo. *Polypedilum prolixipartum* is distinguished from this species by shape of SVo and apical location of distal seta (SVoR less than 0.75), by absence of scutal fossal setae, and shape of apical scale on foretibia.

Description—(n=12)

Coloration: As in *P. vibex* except brown coloration on abdominal tergites II through V very light or faint on some specimens, and apical tarsal segments shaded darker than rest of leg.

Biometrics:	L total 5.35 (4.27-6.19) mm	L thorax 1.43 (1.16-1.78) mm
	L wing 3.04 (2.51-3.34) mm	W wing 0.85 (0.74-0.99) mm

Antenna: AR 1.07 (0.87-1.39) (11). D scape 187 (150-210) (10). L pedicel 130 (120-150) (3). L flagellomeres 1-12 595 (540-680) (11). L flagellomere 13 635 (470-790) (11).

Head: W 608 (520-690). Interocular distance 118 (80-140). HIR 5.15 (4.29-8.00). Vertical S 15.5 (12-21) (10). Clypeal S 19.6 (13-26) (11). Ocelli small, not seen on some specimens. Maxillary palp segment 3 only little longer than segment 4. L maxillary palp segments 1-5: 54 (50-65) (10), 51 (45-56) (10), 194 (173-216), 190 (169-223) (9), 314 (274-362) (9).

Thorax: Dorsal edge of antepronotum moderately produced anteriorly, exceeding anterior margin of scutum in some specimens. Antepronotal campaniform sensillum and postpronotal depression as in *P. fallax*. Antepronotum without setae. Humeral pit present. Acrostichal S 23 (1), in 2 rows, staggered anteriorly but paired posteriorly. Scutal fossal S absent. Some specimens with anterior most dorsocentral

S located well forward, but posterior to imaginary line extended through anterior end of prescutoscutal suture (Figure 24). Dorsocentral S 23.9 (13-29) (11), uniserial or irregularly uniserial anteriorly, irregularly biserial posteriorly. Prealar S 8.7 (7-12). Scutellar S 27.1 (19-35) (11), in 2 or 3 usually irregular rows; posterior row larger.

Wing: VR 1.10 (1.03-1.14). Wing unspotted. Apex of R_{4+5} little to considerably closer to wing apex than apex of M. R_1 and R_{2+3} adjacent apically. Brachiolar S 1 (11). R_1 S 52.6 (39-62). R_{4+5} S 58.9 (44-78). Squamal S 18.4 (16-27) (7).

Legs: Foretibial apical scale rounded apically (Figure 21). Apical spine large and distinct to very small or absent. L foretibial apical scale 47 (34-56). L apical spine (when present) 10 (7-14) (10), may be slightly darkened at tip. L mesotibial apical spine 17 (11-23) (9). L hindtibial apical spine 16 (11-20) (11).

	Fe	**T1**	**Ta1**	**Ta2**	**Ta3**
PI	1073 (860-1290)	1021 (800-1270)	1276 (1110-1520) (9)	798 (680-950) (9)	624 (530-750) (9)
PII	1293 (1090-1510)	1173 (940-1340) (11)	652 (520-740) (11)	381 (300-440) (11)	293 (230-340) (11)
PIII	1393 (1150-1630)	1365 (1110-1610)	968 (790-1110)	521 (420-610)	412 (330-490)

	Ta4	**Ta5**	**LR**	**BV**	**SV**
PI	512 (450-610) (9)	214 (190-230) (8)	1.21 (1.15-1.26) (9)	1.61 (1.56-1.74) (8)	1.69 (1.60-1.79) (9)
PII	179 (130-200) (11)	120 (90-140) (11)	0.55 (0.53-0.57) (11)	3.24 (3.07-3.42) (11)	3.83 (3.63-3.97) (11)
PIII	235 (210-260) (11)	135 (120-160) (11)	0.71 (0.67-0.73)	2.88 (2.74-3.00) (11)	2.87 (2.76-3.04)

Hypopygium: (Figure 125). Tergite IX S 7.3 (3-10) (11), located in relatively small middorsal oval area. Anal point straight and tapered, not exceeding apices of IVo, -59 (-41 to -72) (11). SVo stockier than in other species in *fallax* group, usually narrowed apically beyond origin of distal seta. SVo usually terminating in rounded knob. Basal portion of SVo small, dorsal surface sparsely covered with microtrichia. Basal S 2.3 (1-3) (11). LSVo 108 (99-122). Distal seta located in apical half of SVo. Mean SVoR 0.51 (0.29-0.71). Max L distal S 111 (90-135), usually not exceeding apex of IVo. IVo very long and narrow, dorsal surface free of microtrichia except on extreme medioproximal margin. Apical posteriorly directed seta present. L IVo 208 (187-225). W IVo 32 (25-41) (11). IVo dorsal S 21.0 (15-25), moderately well developed. Gonostylus narrow, medial margin nearly straight, lateral margin evenly rounded. Medial S 7.3 (4-11), in straight and regular row in specimens with few medial setae; in irregular or partially double row in specimens with more setae. Max L medial S 98 (72-113), often reaching the midline. Max L apical S 65 (56-77) (6). L gonostylus 217 (200-243). GsR 3.85 (3.25-4.49). HR 2.30 (2.09-2.52). L posterior lobe gonocoxite 94 (88-101). HV 2.52 (2.30-2.83). Phallapodeme straight, moderately to distinctly expanded at medial end.

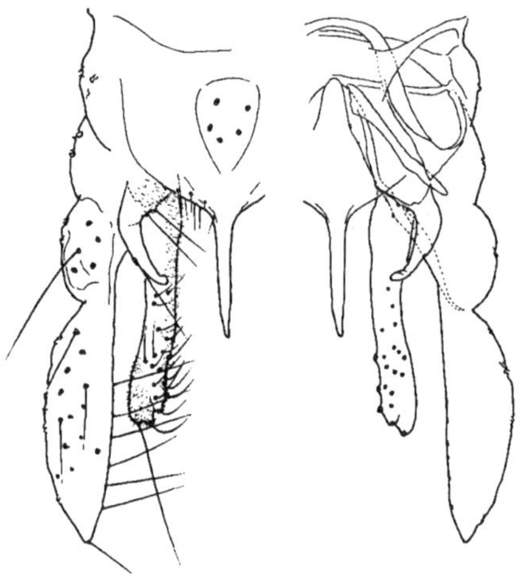

Figure 125. *P. prolixipartum* sp. nov., hypopygium.

Type locality: Holotype male, **Wyoming**, on S fork of Clear Creek, South Fork campground, 16 mi. WSW Buffalo, 24 Aug. 1967, D. Hansen, attracted to illuminated sheet (UMn).

Pupa

Diagnosis—Similar to *P. nubeculosum*, but distinguished from this species by posterior shagreen spinule patches joined medially on tergites III through VI. Distinguished from *P. fallax* by generally sparser shagreen pattern and large fenestrations.

Description—(n=2; males)

Biometrics: L total 5.80 (l) mm L cephalothorax 1.30 (l) mm

Coloration: Typical.

Abdomen: (Figure 126). Pedes spurii B weakly developed. Pedes spurii A present. Row spinules on posterolateral margins of segments V and VI as in *P. fallax* but smaller. Hooks slightly larger than anterior spinules on tergite III. Shagreen spinules small, concentrated in mid portion of tergites II through VI, with large fenestrations. Intersegmental spinules and spinules on tergites VII and VIII as in *P. fallax*. Lateral taeniae 3 on segments V and VI; 4 on segments VII and VIII. Anal spur large, with many small lateral accessory points. Anal taeniae 44 (43-45).

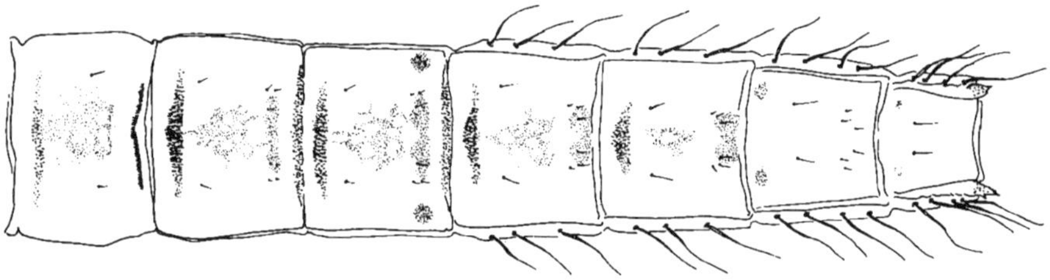

Figure 126. *P. prolixipartum* sp. nov., pupal abdomen.

Remarks

All the remaining specimens examined, except two from Buttle Lake, British Columbia, are designated as paratypes. This species is widely distributed in the western part of North America. It is named for the very long inferior volsellae.

Material Examined

—Reared Specimens

British Columbia, Buttle Lake, 14 Aug. 1964, D.C. Sinclair, 3 males with pupae only, (CNC).

—Adult Specimens

Alaska, Seward Hwy., Bird Creek to Anchorage, 2 Aug. 1964, K. Sommerman, jeep trap, 2 males (USNM); Girdwood Hwy., Anchorage, 9-14 July 1964, K. Sommerman, jeep trap, 1 male, (USNM); Palmer, June 1964, K. Sommerman, jeep trap, 1 male, (USNM). **California**, Leavitt Creek;, Sonora Pass, elv. 8000 ft. Mono Co., 18 July 1968, R.A. Hellenthal, light trap 1 male (UMn); Leavitt Meadow, elv. 7200 ft., Mono Co., 12 Aug. 1963, H.B. Leech, flight trap, 1 male (CalAS); Davis, 9 March 1941, ex collection C.D. and A.H. Michener, 1 male, (UKan). **Colorado**, 27 mi. SE Walden, Jackson Co., 13 and 15 Aug. 1963, N.&B. Marston, Malaise trap, 3 males, (KanSU). **Nevada**, Sweetwater Creek, Toiyabe Nat. Forest, Lyon Co., 3 Sept. 1968, B.S. Cheary, black light, 1 male, (UCR). New Mexico, Rio Grande, Taos Co., 6 July 1953, W.W. Wirth, 6 males, (USNM). **Wyoming**, S fork of Clear Creek, South Fork campground, 16 mi. WSW Buffalo, 23, 24, 27, 28 Aug. 1967, D. Hansen, 5 males, (UMn). Records from March and June through September.

Polypedilum (*U.*) *sulaceps* Townes 1945

Polypedilum sulaceps Townes, 1945: 58; Sublette, 1960: 208; Oyewo and Sæther, 1998.

Adult Male

Diagnosis—This species is close to *P. flavum*. Distinguished from that species by its darker color. Lighter specimens of *P. sulaceps* could be confused with dark *P. flavum*, but even light specimens have a uniformly light brown thorax, whereas for *P. flavum* only the thoracic vittae and postnotum are light brown in dark specimens. In general, LR less than 1.60, SV (foreleg) greater than 1.50, and HR less than 2.90 help separate this species from *P. flavum*. Western in distribution.

Description—(n=5)

Coloration: Head, pedicel, thorax, and legs light brown to brown. Flagellum and abdominal sternites light brown. Haltere stramineous. Abdominal tergites light brown to brown except posterior 1/4 of tergites II through VII which are stramineous to light brown. In light specimens abdominal tergites uniformly stramineous to light brown, wing veins light brown.

Biometrics:	L total 3.80 (3.55-4.06) mm	L thorax 1.10 (0.98-1.21) mm
	L wing 1.96 (1.88-2.12) mm	W wing 0.57 (0.56-0.58) (4) mm

Antenna: AR 1.58 (1.28-1.78). D scape 156 (150-160). L pedicel 120 (120-120) (3). L flagellomeres 1-12 390 (370-430). L flagellomere 13 614 (550-660).

Head: W head 528 (480-550). Interocular distance 110 (90-130). HIR 4.89 (4.23-5.89). Vertical S 11.9 (10-13). Clypeal S 17.6 (15-20). Ocelli present. L maxillary palp 1-5: 42 (36-47), 35 (27-43), 145 (128-169), 156 (140-178), 239 (167-270).

Thorax: Prothorax as in *P. illinoense*. Postpronotal depression small, campaniform sensilla visible in most specimens. Humeral pit large. Acrostichal S 17 (1). Dorsocentral S 15.1 (12-20), uniserial, but irregular posteriorly in large specimens. Prealar S 5.3 (4-6). Scutellar S in 2 rows, anterior row may be irregular, setae in anterior row small to nearly same size as setae in posterior row; anterior row S 9.8 (8-11) (4); posterior row S 11.3 (8-13).

Wing: VR 1.16 (1.11-1.23). Apex of M closer to wing apex than apex of R_{4+5}. R_1 and R_{2+3} adjacent apically. Brachiolar S 1. R_1 S 40.2 (35-45). R_{4+5} S 27.4 (24-36). Squamal S 11.4 (9-14).

Legs: Foretibial apical scale without spine, L scale 31 (27-41). L mesotibial apical spine 18 (16-20). L hindtibial apical spine 20 (18-23).

	Fe	T1	Ta1	Ta2	Ta3
PI	828 (780-860)	618 (560-650)	898 (880-920) (4)	640 (630-670) (4)	480 (470-500) (4)
PII	936 (900-960)	790 (750-820)	485 (460-500) (4)	263 (240-280) (4)	208 (190-220) (4)
PIII	974 (930-1000)	898 (850-940)	698 (650-720) (4)	358 (330-370) (4)	320 (290-340) (4)

	Ta4	Ta5	LR	BV	SV
PI	365 (350-380) (4)	153 (150-160) (4)	1.42 (1.37-1.50) (4)	1.45 (1.43-1.49) (4)	1.64 (1.56-1.69) (4)
PII	133 (120-140) (4)	85 (80-90) (4)	0.61 (0.60-0.62) (4)	3.25 (3.15-3.41) (4)	3.60 (3.54-3.67) (4)
PIII	185 (170-200) (4)	108 (100-120) (4)	0.77 (0.76-0.78) (4)	2.68 (2.61-2.74) (4)	2.72 (2.67-2.75) (4)

Hypopygium: (Figure 127). Tergite IX S 13.0 (8-16), in broad middorsal egg shaped area. Anal point curved ventrally, apical portion usually slightly expanded laterally. Anal point usually about even with apex of IVo, 0 (11 to -16). SVo as in *P. flavum*, except tubercle at origin of distal seta never pronounced. Microtrichia on SVo as in *P. flavum*. Basal S 2.0 (2-2). LSVo 91 (83-99). Mean SVoR 1.09 (0.95-1.31). Max L distal S 92 (79-117). IVo as in *P. flavum*. L IVo 102 (93-110). W IVo 33 (30-36). IVo dorsal S 22.1 (18-30). Gonostylus similar to *P. flavum*, medial margin may be slightly concave. Medial S 7.5 (6-9), very weakly developed. Max L medial S 28 (27-32). Max L apical S 24 (23-25) (3), nearly as long as medial setae. L gonostylus 150 (146-153). GsR variable, 3.57 (3.14-4.25). HR 2.63 (2.52-2.83). L posterior lobe gonocoxite 57 (54-61). HV 2.58 (2.38-2.81). Phallapodeme nearly straight, moderately expanded toward medial end.

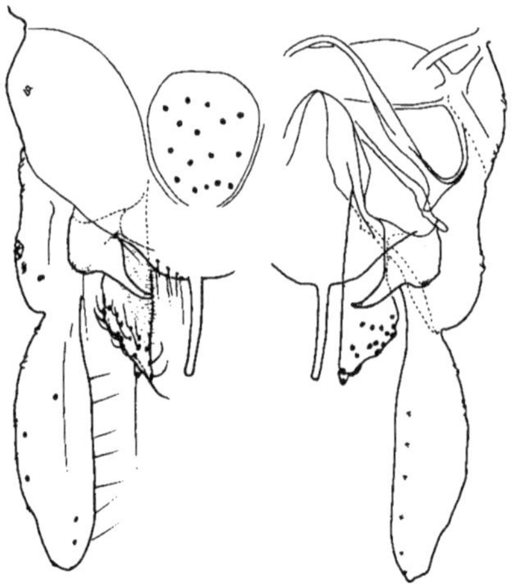

Figure 127. *P. sulaceps* Townes, hypopygium.

Type locality: Holotype male, **Idaho**, Riverdale, 12 Aug. 1934, C.F. Smith, (USNM).

Remarks

As mentioned in the diagnosis, this species is very close to *P. flavum* but it is dark in color and western in distribution. *Polypedilum flavum* has been collected mainly from the eastern two-thirds of the U.S. and Canada.

Material Examined—Adult Specimens

Holotype male, (slide of hypopygium). **Arizona**, N Fork White R. near Roberts Ranch, 9 mi. N White River, 28 June 1961, J. Kingsolver, 2 males, (INH). **California**, Kern Canyon, 15 mi. E Bakersfield, Kern Co., 6 Aug. 1964, R.L. Nelson, light trap, 1 male, (USNM); Richardson Grove State Park, Humboldt Co., 8 July 1944, Shelford, 1 male (hypopygium only), (USNM). **New Mexico**, 5 mi. E Glenwood, Catron Co., 24 June 1953, W.W. Wirth, 4 males (light specimens), (USNM); Cherry Creek, Pinos Altos, 22 June 1953, W.W. Wirth, 2 males (USNM); Cienega campground, 1.9 mi. NW Jct. Hwys. 10 and 44 then 2.6 mi. into campground, Bernalillo Co., 20 June 1968, S. and S. Frommer, at light, 2 males, (UCR). **South Dakota**, 2 mi. N Spearfish, 20 Aug. 1967, D. Hansen at illuminated sheet, 5 males, (UMn). **Utah**, Spanish Fork, 17 Sept. 1964, 3 males, (UMn). **Washington**, Pullman, 18 Sept. 1967, D. Hansen, at lights, 2 males, (UMn). **Wyoming**, Laramie, Albany Co., 25 Aug. 1967, D.J. Ribble, 2 males, (INH). Records from June through September.

Polypedilum (P.) trigonus Townes 1945

Polypedilum trigonus Townes, 1945: 49; Hauber, 1947: 462; Dendy and Sublette, 1959: 513; Beck and Beck, 1959; Beck and Beck, 1964 (larvae and pupae described and figured); Beck and Beck, 1966b: 312; Beck, 1977; Harper and Cloutier, 1979 (biology); Langton, 1984: 253 (figure of pupa); Boesel, 1985: 259; Epler, 1995: 7.95.

Polypedilum trigonum Sublette, 1964: 140.

Adult Male

Diagnosis—Posterolateral corner of abdominal tergites II-V usually with distinct white patch. SVo base large, bearing numerous large microtrichia.

Description—(n=11)

Coloration: Pedicel, head, thorax, abdominal tergites I, VI, VII, and coxae brown. Palps, flagellum, and legs except coxae uniformly light brown to stramineous. Posterolateral corner of tergites II-V usually distinctly lighter than remainder of tergites. Light patches occasionally weak and present only on tergites II and III, or absent altogether in a few specimens. Wing veins stramineous.

Biometrics:	L total 3.47 (2.93-4.08) mm	L thorax 0.95 (0.80-1.05) mm
	L wing 1.60 (1.36-1.84) mm	W wing 0.47 (0.39-0.52) (10) mm

Antenna: AR 1.77 (1.61-1.97) (10). D scape 135 (120-140) (8). L pedicel 110 (100-120) (4). L flagellomeres 1-12 357 (320-400) (10). L flagellomere 13 629 (520-730) (10).

Head: W 503 (460-580) (7). Interocular distance 92 (70-110) (6). HIR 5.47 (4.70-6.57) (6). Clypeal S 20.0 (12-27) (10). Vertical S 12.9 (11-15) (8). Ocelli present. L maxillary palp segments 1-5: 37 (29-47) (9), 109 (97-119) (9), 111 (95-122) (8), 165 (115-198) (9).

Thorax: Dorsal edge of antepronotum not produced anteriorly. Dorsal antepronotal sensilla present on distinct cone. Postpronotal depression without distinct campaniform sensilla. Antepronotum without setae. Dorsocentral S 10.1 (7-13), uniserial. Scutal fossal S absent, Prealar S 4.1 (2-5). Scutellar S 5.9 (4-9) in single row.

Wing: VR 1.19 (1.11-1.25). Wing without spots. R_{4+5} and M apices about equidistant from wing apex, or R_{4+5} sometimes slightly closer. R_1 and R_{2+3} distinctly separated apically. Brachiolar S 1. R_1 S 26.9 (20-31). R_{4+5} S 17.4 (14-24). Squamal S 7.1 (5-10).

Legs: Foretibial apical scale tapered apically to small spine; spine not darkened at tip. L foretibial scale 39 (34-43). L spine 9 (5-16). L mesotibial apical spine 29 (23-34) (8). L hindtibial apical spine 33 (20-41).

	Fe	Ti	Ta1	Ta2	Ta3
PI	726 (610-820)	508 (410-580)	861 (740-1000) (10)	579 (480-690) (10)	407 (320-480) (10)
PII	783 (630-890)	654 (530-740)	434 (390-530) (10)	231 (210-270) (10)	169 (150-200) (10)
PIII	857 (660-970)	755 (620-860)	614 (530-690) (9)	343 (300-380) (9)	300 (260-350) (9)

	Ta4	Ta5	LR	BV	SV
PI	295 (240-340) (10)	137 (110-170) (10)	1.61 (1.43-1.81) (10)	1.50 (1.43-1.59) (10)	1.46 (1.36-1.58) (10)
PII	114 (100-130) (10)	72 (70-80) (10)	0.59 (0.54-0.68) (10)	3.24 (3.07-3.37) (10)	3.40 (3.06-3.66) (10)
PIII	180 (160-210) (9)	87 (70-100) (9)	0.81 (0.76-0.86) (9)	2.46 (2.36-2.55) (9)	2.66 (2.47-2.82) (9)

Hypopygium: (Figures 128, 129). Tergite IX S 6.7 (5-8) (10), located in a light oval-shaped area. Apex of tergite IX tapered to base of anal point as in Figure 128, or rounded at base of anal point in some specimens (Figure 129). Setae at base of anal point large. Anal point tapered slightly from base to apex; apex usually exceeding apices of IVo, 11 (-9 to 36) (9). SVo base large, rounded medially, and baring numerous microtrichia varying from small to large. Group of microtrichia also at base of distal seta. SVo basal S 3.3 (2-4). LSVo 87 (72-104) (10). Distal seta placed near base, mean SVoR 2.11 (1.00-2.78) (10). Max L distal S 69 (56-90) (10). IVo generally long and straight, prominent posteriorly directed apical seta present. L IVo 103 (90-125). W IVo 21 (17-24). IVo dorsal S 9 (6-12). Gonostylus with rounded lateral margin. L gonostylus 149 (113-180). GsR 3.57 (2.94-4.00). Medial S 5.5 (4-7) (10). Medial setae moderately well developed. Max L medial S 57 (47-65). Max L apical S 31 (27-36) (9). L posterior lobe gonocoxite 60 (50-68). HR 2.51 (2.14-2.96). HV 2.37 (2.16-2.82). Phallapodeme usually expanded midway along its length.

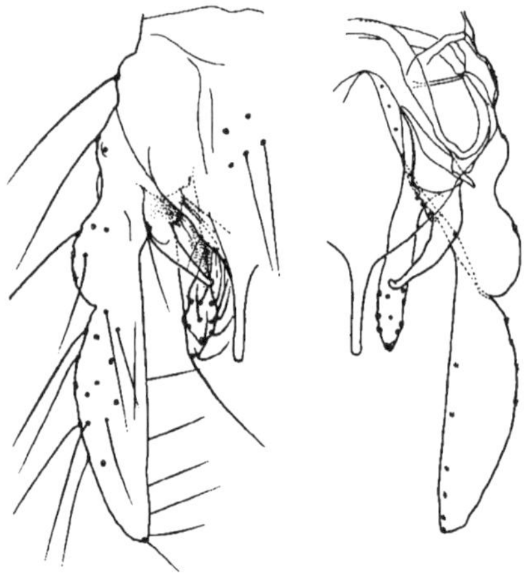

128

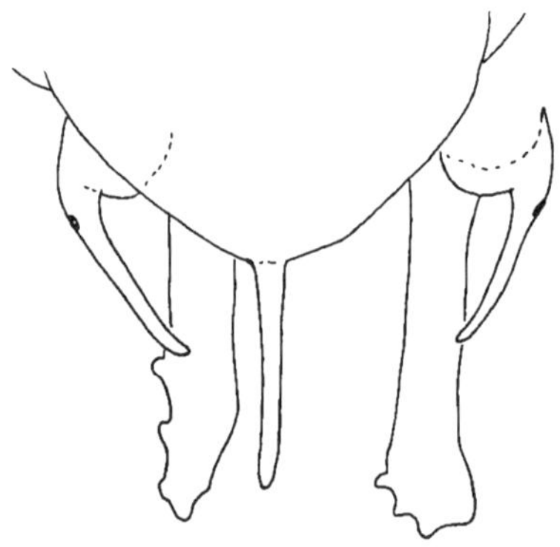

129

Figures 128-129. *P. trigonus* Townes: 128-hypopygium; 129-variation in shape of tergite IX.

Type locality: New York, bank of cove on Hudson River just south of Hudson, 31 Aug. 1936, H. K. Townes, male, (HKT).

Larva

Diagnosis—Distance between ventromental plates small compared to width of ventromental plates, VmPSR 3.63 (3.06-4.40). Lauterborn organs large, extending nearly to apex of third antennal segment.

Description—(n=9; 7 males, 2 females)

Live color: Red

Biometrics: L total 6.04 (5.76-6.31) (2) mm L head capsule 194 (153-223)
L anal gills (approximate) 150 (130-180) (2), with one or two constrictions.

Mentum: (Figure 130). W 103 (90-113) (6), with 16 teeth. Median 2 and second lateral teeth large, third, fourth, and sixth laterals smaller, subequal. Fifth laterals larger and first lateral and penultimate lateral teeth smaller than the three proceeding teeth. Last lateral tooth very small. Teeth uniformly dark, and bluntly pointed. Most specimens with 14 distinct longitudinal dark lines on posterior part of mentum.

Ventromental plates: W 100 (79-117). L 41 (34-50). Medial extensions narrowed to very sharp point. Striations numerous (about 35) and distinct except in most medial part of each ventromental plate. Posterior lobe weakly developed. VmPR 2.45 (2.33-2.58). Distance between ventromental plates 28 (20-36) (8). VmPSR 3.63 (3.06-4.40) (8).

130

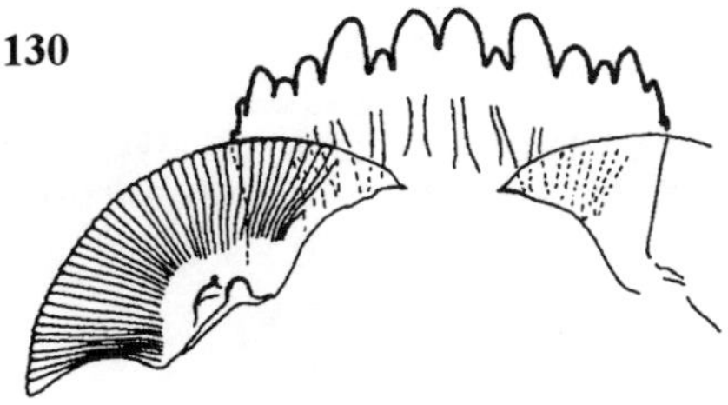

131

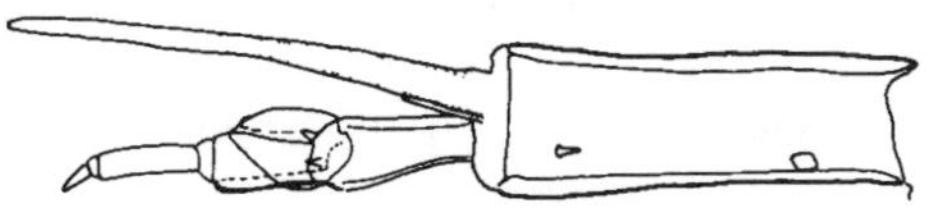

Figures 130-131. *P. trigonus* Townes: 130-mentum; 131-antenna.

Antenna: (Figure 131). L segments 1-5: 52 (45-56), 19 (16-23) (8), 12 (11-14), 11 (9-14), 4 (3-5). AR 0.88 (0.76-0.93) (9). L primary blade (approximately) 50 (43-56) (7), exceeding apex of segment 5. L secondary blade 10 (1). L PS on segment 2 11 (11-12) (2). Lauterborn organs large, leaf shaped, L 8 (5-11) (7). Small triangular structures present on apex of segment 2. ROR 3.80 (3.33-4.40) (5). Segment 1 with 1 sense pit located near apex.

Labrum-epipharynx: Labial setae I-IV as in *P. laetum* except filaments on S II difficult to see. Premandibles bifid. PE in 3 parts, lateral parts with 2-4 teeth, medial part with 3-4 teeth.

Mandible: L 125 (104-149), with 5 teeth as in *P. laetum*. Anteromedial margin with 3 very sharp serrations. Seta subdentalis tapered to a point, L seta subdentalis 21 (18-23) (6). PMa with 4 to 6 filaments.

Maxilla: Similar to *P. nubeculosum*, lacinia prominent, somewhat beak-like in appearance. L lacinia 52 (45-61) (8). Maxillary palp with 3 segments, L segment 1 21 (14-25) (5).

Pupa

Diagnosis—Shagreen spinules on tergites II-VI concentrated near medial part of each tergite, spinules on tergites II and III very small and sparse. Intersegmental spinules absent between segments III and IV.

Description—(n=9; 7 males, 2 females)

Biometrics: L total 4.22 (3.50-5.00) mm

Coloration: Typical.

Abdomen: (Figure 132). Pedes spurii B on segments I and II well developed. Pedes spurii A on segment IV weak to absent. Hooks on tergite II about same size as spinules on anterior margins of tergites II-VI. Intersegmental spinules absent between segments III and IV. Intersegmental spinules between segments IV and V pointing anteriorly, forming broad band on apical portion of segment IV. Shagreen spinules on tergites II and III very small and sparse medially; larger and more extensive medially on tergites IV-VI. Shagreen spinules with large fenestrations on tergites IV-VI. Anterolateral spinules large and distinct on tergite VII, very small and sparse on tergite VIII. Lateral taeniae as in *P. nubeculosum*. Anal spur on segment VIII with 1 large and 1 to 2 smaller apical accessory points. Anal taeniae 28.9 (21-42) on each anal lobe.

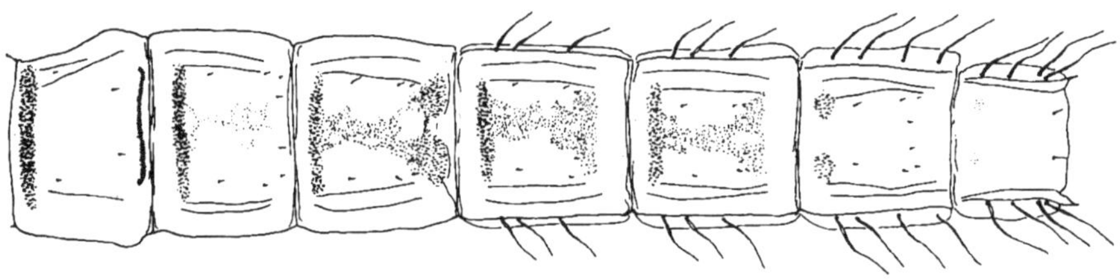

Figures 132. *P. trigonus* Townes, pupal abdomen.

Remarks

A group of specimens from Morrison Field, West Palm Beach, Florida (JES) lacks the posterolateral light patches on the abdominal tergites and the mean values of some parameters differ from *P. trigonus* as described above. However, they generally fall within the range of variation of *P. trigonus* and species recognition does not appear to be justified.

Biology

We have reared this species from larvae collected from cells in thin, partially rotted *Typha* leaves from a small artificial lake in Missouri (Reed Lake), and from the same habitat in a pond in Minnesota (marshy pond, Shoreview). The larvae are usually found near the broken ends of the leaves where they have gained access to the interior of the leaf. Beck (1977) summarized the known environmental requirements of this species. Harper and Cloutier (1979) report *P. trigonus* in small numbers from a few stations in the L'Achigan River, Quebec with moderate to fast moving water. Epler (1995) indicates this is one of the more common species of the northern Everglades, Florida.

Material Examined

—Reared Material

Florida, Boliek's Pond, Leon Co., 15 Oct. 1963, 15 males, 10 April 1963, 2 males, 30 June 1964, 4 males, 11 May 1965, 2 males, W.M. Beck, Jr. (Beck); ditches, State Road No. 210, St. Johns Co., 23 March 1966, 1 male, 17 Nov. 1967. 1 male, W.M. Beck, Jr. (Beck); Lake Miccosuckee, Jefferson Co., 4 June, 1963, 1 male, 15 Jan. 1965, 2 males, 23 July 1965, 1 male, W.M. Beck, Jr. (Beck); Rayonier Ditch, Flager Co., 5 Sept. 1964, 1 male, 23 March 1966, 1 male, W.M. Beck, Jr. (Beck); Roadside ditch in Green Swamp, Polk Co., 16 March 1965, 1 male, W.M. Beck, Jr. (Beck); Expressway Creek, Duval Co., 21 April 1965, 1 male, W.M. Beck, Jr. (Beck); no date, 3 males (Beck). **Minnesota**, marshy pond, Shoreview, Ramsey Co., 45°05'N 93°08'W, 10 Aug. 1971, 1 male, 31 May 1969, 2 females, D.E. Maschwitz (UMn). **Missouri**, Reed Lake, 3 miles S.E. Wappapello, Staddard Co., 12 April 1969, D.E. Maschwitz, 2 males, 1 female, (UMn). Records from January and March through November.

—Adult Specimens

Paratype male, **Florida**, Morrison Field, West Palm Beach, 5 Oct. 1942, D.E. Hardy, light trap (USNM). Two paratype males (slides of hypopygia), no data (USNM), plus about 175 specimens from **Alabama**, **Arkansas**, **California**, **Connecticut**, **Florida**, **Illinois**, **Indiana**, **Iowa**, **Kentucky**, **Louisiana**, **Maryland**, **Massachusetts**, **Michigan**, **Minnesota**, **Montana**, **New York**, **Pennsylvania**, **Virginia**, **Wisconsin**, **Wyoming**, and **Ontario**; from March through October and December.

Polypedilum (P.) tuberculum New Species

Adult Male

Diagnosis—Similar to *P. castornama* and *P. laetum*. Base of SVo with distinct posteroventral tubercle, differing from that found on *P. castornama* by not curving medially and by bearing medially directed setae. All basal setae borne on tubercle, and often set on small secondary tubercles. Wing markings very weak or absent.

Description—(n=8)

Coloration: Pedicel, thorax, and abdominal tergites brown to dark brown, Legs light to dark brown, apices of femora darker on some specimens. Flagellomeres, haltere, and palps brown, abdominal sternites light brown, wing veins stramineous.

Biometrics:	L total 4.14 (3.87-4.39) (6) mm	L thorax 1.12 (1.01-1.20) (6) mm
	L wing 2.32 (2.17-2.50) (7).mm	W wing 0.69 (0.64-0.75) (7) mm

Antenna: AR 1.33 (1.27-1.40) (6). D scape 167 (160-180) (6). L pedicel 110 (1). L flagellomeres 1-12 473 (450-520). L flagellomere 13 628 (590-660).

Head: W 554 (540-570) (7). Interocular distance 109 (100-120) (7). HIR 5.12 (4.75-5.60) (7). Vertical S 14.5 (12-19). Clypeal S 12.6 (10-14). Ocelli small to very small. L maxillary palp segments 1-5: 45 (41-50), 32 (25-41), 147 (133-164), 153 (142-162), 229 (191-252).

Thorax: Postpronotal depression as in *P. laetum*. Antepronotum without setae. Dorsocentral S 17.4 (15-21), uniserial with 2 or more additional setae midway along length of setal row; posterior portion of setal row biserial in some specimens; anterior most dorsocentral seta located well forward in some specimens. Scutal fossal S absent. Prealar S 4.9 (4-6). Scutellar S 18.1 (16-22), biserial with anterior row having as few as 2 setae, setae of 2 rows about same size.

Wing: VR 1.10 (1.06-1.14) (6). Wing spots absent or 2 very faint spots at Fr and fork of M_{3+4} and Cu_1 on some specimens. Apex of M vein closer to wing apex than apex of R_{4+5}. R_1 and R_{2+3} usually slightly separated apically. Brachiolar S 1. R_1 S 31.6 (29-35) (7). R_{4+5} S 27.9 (24-36) (7). Squamal S 15.7 (13-20).

Legs: Foretibial apical scale with spine; spine usually darkened at tip. L scale 41 (38-43) (7). L spine 12 (11-14) (7). L mesotibial apical spine 15 (11-18) (7). L hindtibial apical spine 15 (11-20).

	Fe	T1	Ta1	Ta2	Ta3
PI	884 (810-930)	734 (660-780)	938 (880-1000) (5)	588 (570-630) (5)	452 (430-480) (5)
PII	996 (930-1020)	846 (790-890)	480 (430-500)	305 (280-310)	235 (200-280)
PIII	1141 (1100-1170)	1003 (950-1040)	744 (690-770) (7)	410 (380-420) (7)	341 (310-360) (7)

	Ta4	Ta5	LR	BV	SV
PI	334 (310-370) (5)	154 (150-160) (5)	1.27 (1.18-1.33) (5)	1.68 (1.63-1.75) (5)	1.74 (1.64-1.88) (5)

PII	143	100	0.57	2.99	3.85
	(130-150)	(90-110)	(0.54-0.59)	(2.86-3.16)	(3.74-4.02)
PIII	199	113	0.74	2.73	2.90
	(180-210)	(100-120)	(0.70-0.78)	(2.64-2.85)	(2.80-3.01)
	(7)	(7)	(7)	(7)	(7)

Hypopygium: (Figures 133, 134). Tergite IX S 6.3 (2-8). Apex of tergite IX at base of anal point rounded; slightly truncate in one specimen. Anal point tapered slightly from base to apex, shorter than IVo, -11 (0 to -29) (7). SVo similar to *P. castornama* except that posteroventral tubercle directed posteriorly and setae on tubercle directed medially, perpendicular to long axis of abdomen. These setae often set on small secondary tubercles. SVo basal S 2.3 (2-3), all borne on tubercle. Distal portion of SVo curved toward anal point but often not as long as in *P. laetum* and *P. castornama*. Distal seta moderately long but very fine, max L 75 (65-90) (7). LSVo 83 (74-95). Mean SVoR 1.16 (0.86-1.67). IVo similar to *P. laetum*. L IVo 144 (131-155). W IVo 29 (27-32). IVo dorsal S 12.1 (9-17). Gonostylus as in *P. castornama* except sometimes narrower and medial setae usually not as numerous. Medial S 5.4 (5-6) (7). Max L medial S 66 (59-70) (6). L gonostylus 158 (146-171). GsR 3.22 (2.96-3.50) (7). L posterior lobe gonocoxite 71 (63-81), HR 2.23 (1.89-2.40). HV 2.61 (2.44-2.78). Microtrichia on basal plate well developed. Phallapodeme similar to *P. laetum*.

133

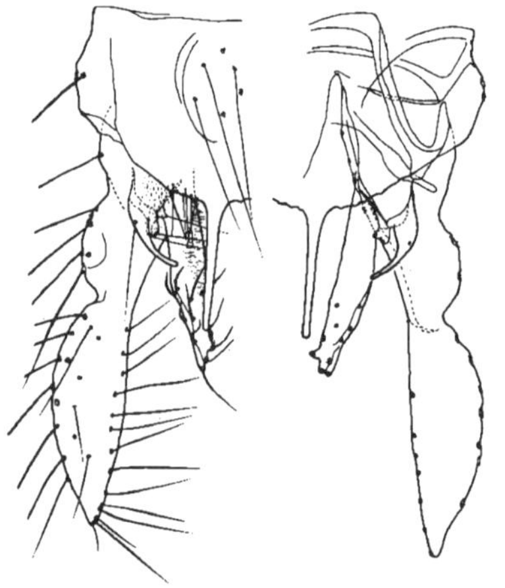

134

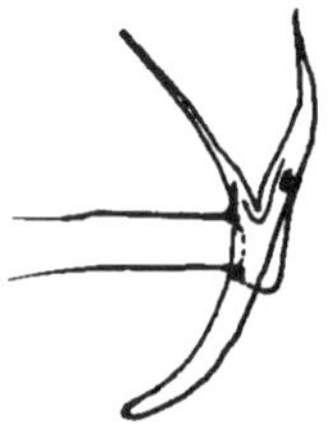

Figures 133-134. *P. tuberculum* sp. nov., 133-hypopygium; 134-variation of superior volsella.

Type locality: Holotype male, **Minnesota**, 2 mi. N. Hovland on U.S. Highway 61, Cook Co., 10 Aug. 1970, leg. E.F. Cook, New Jersey mosquito light trap, (UMn).

Larva

Diagnosis—Very close to *P. laetum*. The following characters should separate most specimens: Ring organ located more distally than in *P. laetum*, ROR 2.79. Antennal segment 1 apparently without basal sense pit. Antennal segments 3 and 4 usually shorter than in *P. laetum*.

Description—(n=5; 4 males, 1 female)

Biometrics: L head capsule 153 (149-155) (4)

Mentum: (Figure 135). W 101 (97-106) (4), with 16 teeth; last lateral teeth so small that mentum may be considered to have 14. Teeth generally even in height and color. Size of teeth as in *P. laetum*, but apices of teeth evenly rounded, or, in some specimens, median 2 and first lateral teeth somewhat flattened.
Ventromental plates: W 113 (108-117) (4). L 52 (45-59) (4). General shape similar to *P. laetum*, but

posterior lobe appearing little more strongly developed in most specimens; striations distinct along entire width of ventromental plate. VmPR 2.22 (1.92-2.50) (4). Distance between ventromental plates 52 (50-56) (4). VmPSR 2.17 (1.96-2.27).

135

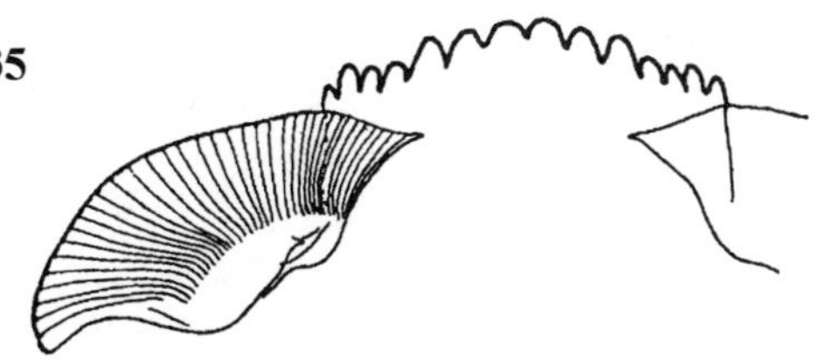

136

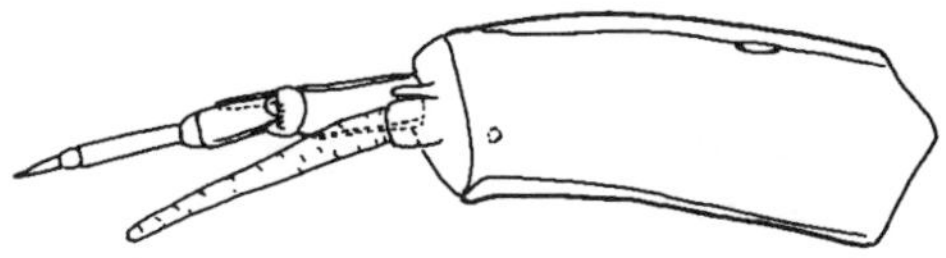

Figures 135-136. *P. tuberculum* sp. nov.: 135-mentum; 136-antenna.

Antenna: (Figure 136). L segments 1-5: 59 (58-61) (2), 17 (17-18) (2), 10 (10-11) (2), 14 (14-14) (2), 5 (5-6) (2). AR 0.78 (0.77-0.79) (2). L primary blade (approx.) 37 (35-39) (2). L secondary blade 7 (1). L PS on segment 2 7 (7-8) (2). Lauterborn organ very obscure, not visible on some specimens. Two triangular structures present on apex of segment 2. ROR 2.79 (2.67-2.95) (3). Segment 1 apparently having only 1 sense pit.

Labrum-epipharynx: Labral S I-IV as in *P. laetum*. Premandibles bifid, with 1 tooth almost always longer than the other. PE in 3 parts, lateral parts with 3-4 teeth, medial part with 3 teeth.

Mandible: L 120 (117-124) (4), with 5 teeth as in *P. laetum*. Medial margin with 2 sharply pointed serrations. L seta subdentalis 20 (16-23) (3), pointed apically. PMa with about 4 filaments.

Maxilla: Similar to *P. nubeculosum*. L lacinia 40 (34-43). Maxillary palp with 3 segments, L segment 1 19 (1).

Pupa

Diagnosis—Close to *P. laetum* but distinguished by following: Smaller shagreen spinules. Shagreen spinules distinctly smaller than hooks on posterior margin of tergite II. Shagreen pattern on tergites II through VI triangular shaped. Intersegmental spinules small, about same size as shagreen spinules.

Description—(n=4; 3 males, 1 female)

Biometrics: L total 4.4 (4.3-4.5) mm (3)

Coloration: Typical.

Abdomen: (Figure 137). Pedes spurii B on segment I moderately developed, those on segment II small. Pedes spurii A moderately developed. Hooks on tergite II smaller than spinules on anterior margins of tergites II-VI. Intersegmental spinules and shagreen spinules small, about same size. Intersegmental spinules multiserial, pointing both anteriorly and posteriorly. Shagreen spinule pattern roughly triangular in shape on tergites II through VI, with large fenestrations. Tergites VII and VIII with anterolateral patches of very small spinules. Lateral taeniae as in *P. nubeculosum*. Anal spur on segment VIII with 1 to 5 apical points and many smaller lateral accessory points. Anal taeniae 23.9 (22-26) on each anal lobe.

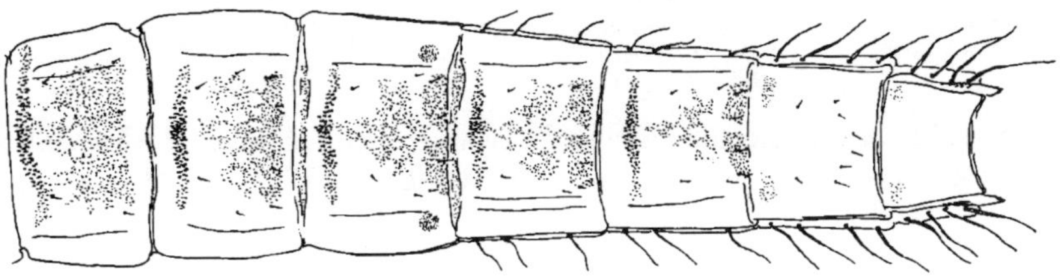

Figure 137. *P. tuberculum* sp. nov., pupal abdomen.

Remarks

Paratypes are designated as indicated in material examined. The species is named for the tubercle on the SVo base. Pinder and Reiss (1983: 423) illustrate the mentum and larval antenna.

Biology

Larvae of this species were found in bottom sediment samples taken from Lake Superior at depths ranging from 4 to 20 meters. Bottom substrates ranged from silty sand with large amounts of organic debris to coarse sand and gravel.

Material Examined

—Reared Specimens

New Mexico, Animas River at Aztec, San Juan Co., 17-24 April 1968, (JES), 3 male and 3 female paratypes, 1 male, 2 females.

—Adult Specimens

Alaska, Matanuska Eklutna Hwy., 22 June 1964, K.M. Sommerman, jeep trap, 1 male, (USNM); Seward Hwy., Bird Creek to Anchorage, 2 Aug. 1964, 1 male; Anchorage, Seward Hwy., 25 Aug. 1964, K.M. Sommerman, jeep trap, 1 male paratype (USNM). **Minnesota**, 2 mi. N. Hovland on U.S. Highway 61, Cook Co., July and August 1970, 26 male paratypes; July through September 1970, 13 males, leg. E.F. Cook, New Jersey mosquito light trap, (UMn); Duluth, St. Louis Co., 12 Sept 1969, 20 Sept 1969, 23 July 1970, 3 male paratypes, Sept 1969, 7 males, leg. E.F. Cook, New Jersey mosquito light trap, (UMn). **New Mexico**, Questa Ranger Station, 27 July 1967, W.R. Atchley, light trap, 1 male paratype, (JES); Cienega campground, 1.9 mi. N.W. junction Highways 10 and 44, then 2.6 mi. into campground, Bernalillo Co., 20 June 1968, S. and S. Frommer, at light, one male paratype, (UCR). **Wisconsin**, Sturgeon Bay, Door Co., 2 Aug. 1961, E.F. Legner, 1 male (UW). **Wyoming**, Grand Teton National Park, Colter Bay, Jackson Lake, Teton Co., 4 July 1966, W. M. Griffith, 1 male, 1 male paratype, 25 July 1967, 5 males, 1 male paratype, (INH). **British Columbia**, Silver Creek, Hope, 3 July 1968, W.W. Wirth, light trap, 1 male paratype, (USNM); Revelstoke, 1 July 1968, W.W. Wirth, light trap, 2 male paratypes, (USNM). **Ontario**, Agawa Bay, Lake Superior Provincial Park, 23 July 1970, W.E. LaBerge, 1 male, 1 male paratype, (INH); Southampton (Lake Huron), Bruce Co., 17 Sept. 1967, D.W. Webb, 3 male paratypes, (INH). Records from June through September.

Polypedilum (*P.*) *vibex* Townes 1945

Polypedilum vibex Townes, 1945: 52; Townes, 1952: 51; Beck and Beck, 1959: 92; Boesel, 1985: 259; Harper and Cloutier, 1979 (biology).

Adult Male

Diagnosis—Forefemur stramineous, abdominal tergites II through V with anterior part brown, posterior part stramineous; extent of brown portion increasing posteriorly from one-fourth of tergite II to one-half of tergite V. Scutal fossal S present. IVo long. SVo and foretibial apical scale as in *P. fallax*.

Description—(n=2)

Coloration: Vertex of head, pedicel, thorax (except scutellum), portions of coxae, and abdominal tergites VI, VII, and VIII brown. Haltere, scutellum, maxillary palps, flagellum, legs, hypopygium, and wing veins stramineous to light brown. Abdominal tergite I stramineous, tergites II through V as described in diagnosis.

Biometrics:	L total 6.19 (5.81-6.58) mm	L thorax 1.77 (1.69-1.84) mm
	L wing 3.08 (2.88-3.28) mm	W wing 0.89 (0.81-0.97) mm

Antenna: AR 1.36 (1.31-1.41). D scape 220 (220-220). L pedicel 150 (1). L flagellomeres 1-12 745 (740-750). L flagellomere 13 1000 (930-1060).

Head: W 750 (720-780). Interocular distance 90 (90-90). HIR 8.34 (8.00-8.67). Vertical S 21.0 (19-22). Clypeal S 43.0 (41-45). Ocelli small. L maxillary palp segments 1-5: 61 (1), 43 (1), 216 (209-223), 195 (187-203), 338 (306-369).

Thorax: Antepronotum and thoracic sensilla similar to *P. fallax*. Antepronotum without setae; campaniform sensilla in postpronotal depression visible but not as distinct as in *P. fallax*. Scutal fossal S 4.3 (3-6). Dorsocentral S 30 (28-33), uniserial to irregularly biserial. Prealar S 14.0 (10-18). Scutellar S approximately 39 (33-45) in 3 irregular rows.

Wing: VR 1.19 (1.18-1.20). Wing unspotted. Apex of R_{4+5} closer to wing apex than apex of M. R_1 and R_{2+3} adjacent or separated by slight gap apically. Brachiolar S 1. R_1 S 86.5 (81-92). R_{4+5} S 106.0 (90-122). Squamal S approximately 30.

Legs: Shape of foretibial apical scale as in *P. fallax*. L (total) 84 (68-86). L mesotibial apical spine 27 (25-29). L hindtibial apical spine 33 (27-38).

	Fe	**T1**	**Ta1**	**Ta2**	**Ta3**
PI	1390 (1330-1450)	1105 (1050-1160)	177 (1)	117 (1)	90 (1)
PII	1695 (1620-1770)	1400 (1330-1470)	825 (800-850)	530 (510-550)	435 (410-460)
PIII	1770 (1720-1820)	1575 (1510-1640)	1250 (1)	740 (1)	620 (1)

	Ta4	**Ta5**	**LR**	**BV**	**SV**
PI	700 (1)	270 (1)	1.52 (1)	1.45 (1)	1.48 (1)
PII	285 (270-300)	140 (140-140)	0.59 (0.57-0.60)	2.84 (2.83-2.84)	3.78 (3.71-3.84)
PIII	380 (1)	160 (1)	0.77 (1)	2.50 (1)	2.76 (1)

Hypopygium: (Figure 138). Tergite IX S 17, located in middorsal egg-shaped area. Anal point straight, tapered apically, terminating well short of IVo apices, -58 (-45 to -70). SVo similar to *P. fallax*. Basal S 3.5 (3-4). LSVo 154 (146-162). Mean SVoR 0.83 (0.65-0.97). Max L distal S 167 (149-185). IVo very long and narrow. L IVo 225 (214-230). W IVo 38 (36-39). Apical posteriorly directed seta present. IVo dorsal S 25 (23-27). Gonostylus as in *P. fallax*. Medial S 5.3 (5-6), usually in straight row. Max L medial S 112 (99-124). Max L apical S 53 (43-63). L gonostylus 280 (275-286). GsR 4.05 (3.74-4.36). HR 2.15 (2.12-2.18). L posterior lobe gonocoxite 131 (126-135). HV 2.25 (2.21-2.28). Phallapodeme straight, expanded midway along its length and at medial end.

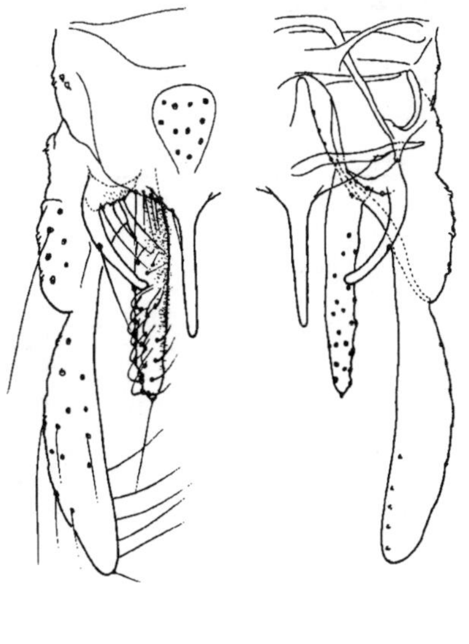

Figure 138. *P. vibex* Townes, hypopygium.

Type locality: Holotype male, **Ontario**, Ottawa, 26 July 1924, C.H. Curran, male, (CNC).

Biology

Harper and Cloutier (1979) report *P. vibex* from L'Achigan River stations with medium to fast flowing water but in very small numbers.

Material Examined—Adult Specimens

Holotype male. Paratype male, same as holotype, (USNM).

Polypedilum (*P.*) *walleyi* Townes 1945

Polypedilum walleyi Townes, 1945: 52; Townes, 1952: 51; Boesel, 1985: 257 (*P. calopterum*, new synonymy); Oliver *et al.*, 1990: 51 (listing of synonymy); Caldwell *et al.*, 1997.

Adult Male

Diagnosis—Forefemur uniformly light brown except apical 0.15 which is stramineous. Abdomen and hypopygium uniformly stramineous. Large species.

Description—(n=l; holotype male)

Coloration: Vertex of head, pedicel, thorax, and forecoxae brown. Flagellum light brown. Legs stramineous except apices of forefemora as noted in diagnosis, and apices of tarsi, which are light brown. Haltere, maxillary palps, wing veins, and abdomen stramineous.

Biometrics:	L total 7.00 mm	L thorax 1.99 mm
	L wing 3.52 mm	W wing 1.04 mm

Antenna: AR 1.46. D scape 250. L pedicel 150. L flagellomeres 1-12 720. L flagellomere 13 1040.

Head: W 860. Interocular distance 140. HIR 6.14. Vertical S approximately 30. Clypeal S 46. Ocelli small. L maxillary palp segments 1-5: 56, 68, 234, 221, 394.

Thorax: Antepronotum and thoracic sensilla similar to *P. fallax*. Humeral pit absent. Antepronotum without setae. Scutal fossal S 2. Dorsocentral S 57, biserial to triserial. Prealar S 23. Scutellar S approximately 46 in 3 irregular rows.

Wing: VR 1.13. Wing unspotted. Apex of R_{4+5} closer to wing apex than apex of M. R_1 and R_{2+3} slightly separated apically. Brachiolar S 1. R_1 S 71. R_{4+5} S 58. Squamal S 25.

Legs: Foretibial apical scale as in *P. fallax*, total L 81. L mesotibial apical spine 32. L hindtibial apical spine 38.

	Fe	**T1**	**Ta1**	**Ta2**	**Ta3**
PI	1640	1400	2050	1320	1050
PII	2030	1690	950	600	450
PIII	2190	1840	1410	780	660
	Ta4	**Ta5**	**LR**	**BV**	**SV**
PI	880	360	1.47	1.41	1.48
PII	300	170	0.56	3.09	3.95
PIII	410	210	0.77	2.66	2.86

Hypopygium: (Figure 139). Tergite IX S 26, located in middorsal oval shaped area. Anal point straight, tapered apically, terminating short of IVo apices, -43. SVo similar to *P. fallax*. Basal S 3.5. LSVo 172 (171-173). Mean SVoR 0.69 (0.62-0.75). Max L distal S 105, ending near apices of IVo. IVo as in *P. fallax*. L IVo 258 (257-259). W IVo 45 (45-45). IVo dorsal S 35.5 (35-36). Gonostylus as in *P. fallax*. Medial S 5.0. Max L medial S 124. Max L apical S 86. L gonostylus 311. GsR 3.36. HR 2.23. L posterior lobe gonocoxite 140. HV 2.27. Basal plate not distinct. Phallapodeme straight, not expanded.

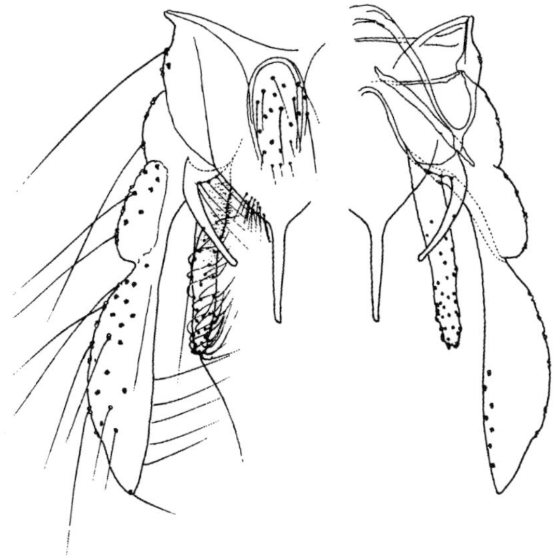

Figure 139. *P. walleyi* Townes, hypopygium.

Type locality: **Quebec**, Aylmer, 14 Aug. 1924, G.S. Walley, male, (CNC).

Remarks

In recognizing *P. fallax* and *P. calopterum* as separate species, Boesel (1985) also concluded that *P. walleyi* is a synonym of *P. calopterum*. Oliver *et al.* (1990) subsequently listed *P. walleyi* as a synonym of *P. calopterum*. However, Caldwell *et al.* (1997) did not accept this synonymy. The description above is based on the holotype alone. We have seen no other specimens which could be identified with certainty as *P. walleyi*. We have not reexamined specimens of these three species, and, therefore, do not recognize the synonymy in this work.

Literature Cited

Albu, Paula. 1963. Chironomidae (Adulte) din Bazinul Sornesului si al viseului (Muntii rodnei). Studii si Cercetari de Biologie, Seria Zoologie 15: 223-236.

___. **1980.** Fam. Chironomidae, Subfam. Chironominae. Fauna Republicii Socialiste Romania, Insecta, Diptera. 11(13): 1-320. [not seen]

Amakye, Joseph S. 1995. *Collartomyia discaudata*, spec. nov. from Ghana, with an emendation of the genus (Insecta, Diptera, Chironomidae). Spixiana 18: 271-275.

Amakye, Joseph and Ole A. Sæther. 1992. The immatures and imagines of the Afrotropical species *Microtendipes lentiginosus* Freeman and *Collartomyia hirsuta* Goetghebuer (Diptera: Chironomidae). Entomologica scandinavica 23: 429-442.

Anthon, H. 1943. Der Kopfbau der Larven einiger nematoceren Dipterenfamilien: Rhyphidae, Trichoceridae, Psychodidae und Ptychopteridae. Spolia Zoologica Musei Hauniensis 3: 1-61.

Ashe, Patrick. 1981. A taxonomic note on the genus *Polypedilum* Kieffer (Diptera: Chironomidae) with an earlier date for the genus and designation of a type-species. Aquatic Insects 3: 57-59.

Ashe, Patrick and Peter S. Cranston. 1991. Family Chironomidae. Pages 113-355 *in* Catalogue of Palaearctic Diptera. Volume 2 Psychodidae—Chironomidae. A. Soos and L. Papp, eds. Akademiai Kiado, Budapest and Elsevier Science Publishing Co. New York, New York. 499 p.

Bause, E. 1913. Die Metamorphose der Gattung *Tanytarsus* und einiger verwandter Tendipedidenarten. Ein Beitrag zur Systematik der Tendipediden. Pages 1-126 *In* Vorarbeiten für eine Monographie der Chironomiden-Metamorphose by August Thienemann. Archiv für Hydrobiologie Supplementband 2. 241 p.

Beck, William M., Jr. 1977. Environmental requirements and pollution tolerance of common freshwater Chironomidae. Environmental Monitoring and Support Laboratory. U.S. Environmental Protection Agency, Cincinnati, Ohio. EPA-600/4-77-024. 261 p.

Beck, Elizabeth C. and William M. Beck, Jr. 1959. A checklist of the Chironomidae (Insecta) of Florida (Diptera: Chironomidae). Bulletin of the Florida State Museum, Biological Sciences 4: 85-96.

___. **1964.** New Chironomidae from Florida (Diptera). Florida Entomologist 47: 201-207.

___. **1966a.** The Chironomidae of Florida: a problem in international taxonomy. Gewässer und Abwässer 41/42: 129-135.

___. **1966b.** Chironomidae (Diptera) of Florida I. Pentaneurini (Tanypodinae). Bulletin of the Florida State Museum, Biological Sciences 10: 305-379.

Berg, Clifford O. 1950. Biology of certain Chironomidae reared from *Potamogeton*. Ecological Monographs 20: 83-101.

Bidawid, Najwa and Ernst J. Fittkau. 1995. Zur Kenntnis der neotropischen Arten der Gattung *Polypedilum* Kieffer, 1912 Teil I (Diptera: Chironomidae). Entomofauna Zeitschrift für Entomologie 16: 465-536.

Bidawid-Kafka, Najwa. 1996. Zur Kenntnis der neotropischen Arten der Gattung *Polypedilum* Kieffer, 1912 Teil II (Diptera: Chironomidae). Entomofauna Zeitschrift für Entomologie 17: 165-240.

Bjørio, Alfred, Hege Vårdal, and Ole A. Sæther. 1997. A preliminary phylogenetic analysis of the subgenus *Tripodura* Townes of the genus *Polypedilum* Kieffer (Diptera: Chironomidae). Abstract of paper presented at the 13th International Symposium on Chironomidae. Freiburg, Germany, 5-9 September, 1997.

Boesel, M.W. 1985. A brief review of the genus *Polypedilum* in Ohio, with keys to known stages of species occurring in northeastern United States (Diptera, Chironomidae). The Ohio Journal of Science 85: 245-262.

Bolton, Michael J. 1991. The identity of Chironomini genus C (Diptera:Chironomidae) in Pinder and Reiss (1986). Enomological News 102: 125-126.

Branch, H.E. 1931. Identification of chironomid egg masses. II. Transactions of the Kansas Academy of Science 34: 151-157.

Brundin, Lars. 1947. Zur Kenntnis der schwedischen Chironomiden. Arkiv för Zoologi 39 A (3): 1-95.

___. **1949.** Chironomiden und andere Bodentiere der südschwedischen Urgebirgsseen. Ein Beitrag zur Kenntnis der Bodenfaunistischen Charakterzüge Schwedischer Oligotropher Seen. Institute of Freshwater Research, Drottningholm. Report No 30. 914 p.

___. **1966.** Transantarctic relationships and their significance, as evidenced by chironomid midges, with a monograph of the subfamilies Podonominae and Aphroteniinae and the austral Heptagyiae. Kungl. Svenska Vetenskapsakad.emiens Handlinger 11 (1): 1-472.

Bryce, Derek. 1960. Studies on the larvae of the British Chironomidae (Diptera) with keys to the Chironominae and Tanypodinae. Transactions of the Society for British Entomology 14: 19-62.

Caldwell, Broughton A., Patrick L. Hudson, David R. Lenat, and David R. Smith. 1997. A revised annotated checklist of the Chironomidae (Insecta: Diptera) of the southeastern United States. Transactions, American Entomological Society 123: 1-53.

Chernovskii, Aleksei A. 1949. Identification of larvae of the midge family Tendipedidae. Opredelitel' Faune SSSR, 31. 186 p. English translation by Dr. E. Lees (K.E. Marshall, ed.), 1961. National Lending Library for Science and Technology. Boston Spa, England. 300 p.

Coffman, William P. 1986. 1. The pupae of Chironomidae (Diptera) of the Holarctic region—keys to subfamilies. Pages 9-11 *in* Chironomidae of the Holarctic region—keys and diagnoses. Part 2. Pupae. Torgny Wiederholm, ed. Entomologica scandinavica Supplement 28. 482 p.

Cook, Edwin F. 1944a. The morphology and musculature of the labrum and clypeus of insects. Microentomology 9: 1-35.

___. **1944b.** On the morphology of the larval head of a species of *Chironomus* (Diptera: Chironomidae). Microentomology 9: 69-77.

___. **1949.** The evolution of the head in the larva of the Diptera. Microentomology 14 (1): 1-57.

Craig, D.A. 1969. The embryogenesis of the larval head of *Simulium venustum* Say (Diptera: Nematocera). Canadian Journal of Zoology 47: 495-503.

Cranston, Peter S. 1989. New species of Chironominae (Diptera: Chironomidae) from Saudi Arabia and the adjacent Middle East. Fauna of Saudi Arabia 10: 225-235.

Cranston, Peter S., Mary E. Dillon, L.Clive V. Pinder, and Friedrich Reiss. 1989. 10. The adult males of Chironominae (Diptera: Chironomidae) of the Holarctic region—keys and diagnoses. Pages 353-502 *in* Chironomidae of the Holarctic region—keys and diagnoses. Part 3. Adult males. Torgny Wiederholm, ed. Entomologica scandinavica Supplement 34. 532 p.

Cranston, Peter S. and D.D. Judd. 1989. Diptera: Fam. Chironomidae of the Arabian peninsula. Fauna of Saudi Arabia 10: 236-289.

Cranston, Peter S. and Friedrich Reiss. 1983. 2. The larvae of Chironominae (Diptera: Chironomidae) of the Holarctic region—key to subfamilies. Pages 11-15 *in* Chironomidae of the Holarctic region keys and diagnoses. Part 1. Larvae. Torgny Wiederholm, ed. Entomologica scanddinavica Supplement 19. 457 p.

Curran, Charles H. 1930. Report on the Diptera collected at the station for the study of Insects, Harriman Interstate Park, New York. Bulletin American Museum of Natural History (1931) 61: 21-115.

Dendy, J.S. and James E. Sublette. 1959. The Chironomidae (=Tendipedidae: Diptera) of Alabama with descriptions of six new species. Annals of the Entomological Society of America 52: 506-519.

Dittmar, Helmut. 1955. Ein Saurlandbach. Untersuchungen an einen Wiesen-Mittelgebirgsbach. Archiv für Hydrobiologie 50: 305-552.

Edwards, Frederick W. 1929. British non-biting midges (Diptera, Chironomidae). Transactions of the Royal Entomological Society of London 77: 279-430.

___. **1931.** Diptera of Patagonia and south Chile, Part II, fascicle 5.—Chironomidae. British Museum (Natural History) 1931: 233-324.

van Emden, F. and Willi Hennig. 1970. 21. Diptera. Pages 130-141 *in* Taxonomist's glossary of genitalia in insects. Soren L. Tuxen, ed. Second revised and enlarged edition. Munksgaard, Copenhagen. 359 p.

Epler, John H. 1995. Identification manual for the larval Chironomidae (Diptera) of Florida. Revised Edition. Florida Department of Environmental Protection. Tallahasee, Florida. 317 p.

Fittkau, Ernst J. 1962. Die Tanypodinae (Diptera: Chironomidae) (Die Tribus Anatopyniini, Macropelopiini und Pentaneurini). Abhandlungen zur Larvalsystematik der Insekten, Nr. 6. 453 p.

___. **1971.** Distribution and ecology of Amazonian chironomids (Diptera). The Canadian Entomologist 103: 407-413.

Fittkau, Ernst J., Dieter Schlee, and Friedrich Reiss. 1967. Chironomidae. Pages 346-381 *in* Limnofauna Europaea. Joachim Illies, ed. Stuttgart, Germany. 474 p.

Freeman, Paul. 1958. A Study of the Chironomidae (Diptera) of Africa south of the Sahara, Part IV. Bulletin of the British Museum (Natural History) Entomology 6: 261-363.

___. **1959.** A study of the New Zealand Chironomidae (Diptera, Nematocera). Bulletin of the British Museum (Natural History) Entomology 7: 393-437.

___. **1961.** The Chironomidae (Diptera) of Australia. Australian Journal of Zoology 9: 611-737.

Frommer, Saul. 1967. Review of the anatomy of adult Chironomidae. California Mosquito Control Association Technical Services Bulletin 1. 40 p.
Gimmerthal, B.A. 1845. Erster Beitrag zu einer künftig zu bearbeitenden Dipterologie Russlands. Familie de Tipularien. Bulletin de la Société Impériale des Naturalistes de Moscou 18: 287-331.
___. **1846.** Zweiter Beitrag zur Dipterologie Russlands. Bulletin de la Société Impériale des Naturalistes de Moscou 19: 3-82.
Goetghebuer, Maurice. 1912. Études sur les Chironomides de Belgique. Académie Royale de Belgique, Mémoires de la classe des sciences 3: 1-26.
___. **1921.** Chironomides de Belgique et spécialement de la zone des Flanders. Mémoires du Musée Royal d'Histoire Naturelle de Belgique 8: 1-211.
___. **1923.** Étude critique des chironomides de la collection Meigen conservée au Museum d'Histoire Naturelle de Paris. 2e note. Bulletin de la Société Entomologie de Belgique 5: 120-129.
___. **1928.** Diptères (Nématocères): Chironomidae. III. Chironomariae. Vol. 18 of Fauna de France. Paris. 174 p.
___. **1931a.** Ceratopogonidae et Chironomidae nouveaux d'Europe. Bulletin & Annales de la Société Royale d'Entomologique de Belgique 71: 211-218.
___. **1931b.** Chironomides paléarctiques (Diptères) conservés au Musée d'Histoire Naturelle de Vienne. Annalen des Naturhistorischen Museums in Wien. 46: 91-115.
___. **1934.** Catalogue des Chironomides de Belgique V.—Chironominae. Bulletin & Annales de la Société Royale d'Entomologique de Belgique 74: 391-405.
___. **1936.** Les Cèratopogonides et les Chironomides de Belgique. Au point de vue hydrobiologique. Bulletin & Annales de la Société Royale d'Entomologique de Belgique 76: 69-76.
___. **1937.** Tendipedidae (Chironomidae). b) Subfamilie Tendipedinae (Chironominae). Pages 49-72 *in* Die Fliegen der palaearktischen Region. Erwin B. Lindner, ed. 13c, Lieferung 109. 49-72 p. und Tafel VIII-XIV.
___. **1938.** Tendipedidae (Chironomidae). b) Subfamilie Tendipedinae (Chironominae). Pages 73-128 *in* Die Fliegen der palaearktischen Region. Erwin B. Lindner, ed. 13c, Lieferung 118. 73-128 p. und Tafel XV-XIX.
___. **1939.** Ceratopogonidae et Chironomidae nouveaux ou peu connus d'Europe (Dixieme note). Bulletin & Annales de la Société Royale d'Entomologique de Belgique 79: 379-389.
Gouin, Francois. 1957. Études sur l'anatomie de la tête larvaire de quelques Chironomidae (Dipt. Nematoc.). I.—Les structures Chironomines. Annales de la Société Entomologique de France 126: 105-138.
___. **1959.** Morphology of the larval head of some Chironomidae (Diptera, Nematocera). Smithsonian Miscellaneous Collections 137: 175-201.
Grodhaus, Gail and George L. Rotramel. 1980. Immature stages of *Polypedilum pedatum excelsius* (Diptera, Chironomidae) from seasonally flooded tree-holes. Acta Universitatis Carolinae—Biologica 1978: 69-76.
Hamilton, Andrew L. 1965. An analysis of a freshwater benthic community with special reference to the Chironomidae. II. Taxonomy of the Chironomidae of Marion Lake, British Columbia. Ph.D. Thesis. Department of Zoology, University of British Columbia. Vancouver, British Columbia. 216 p.
___. **1971.** Zoobenthos of fifteen lakes in the experimental lakes area, northwestern Ontario. Journal of the Fisheries Research Board of Canada 28: 257-263.
Hamilton, Andrew L., Ole A. Sæther, and Donald R. Oliver. 1969. A classification of the Nearctic Chironomidae. Fisheries Research Board of Canada Technical Report No. 124. 42 p.
Hansen, Dean C. and Edwin F. Cook. 1976. Systematics and morphology of the Nearctic species of *Diamesa* Meigen, 1835 (Diptera: Chironomidae). Memoirs of the American Entomological Society Number 30. 203 p.
Harnisch, O. 1922. Zur Kenntnis der Chironomidenfauna der Brassenregion schlesischer Flüsse. Archiv für Hydrobiologie 14: 125-143.
Harper, P.P. and Louise Cloutier. 1979. Chironomini and Pseudochironomini of a Quebec highland stream (Diptera: Chironomidae). Entomologica scandinavica Supplement 10: 81-94.
Harrison, Arthur D. 1996. Chironomidae from Ethiopia, Part 3. Chironomini, with description of a new species (Insecta, Diptera). Spixiana 19: 43-87.
Hauber, U.A. 1947. The Tendipedinae of Iowa (Diptera). The Ameican Midland Naturalist 38: 456-465.
Hennig, Willi. 1948. Die larvenformen der Dipteren. Eine Übersicht über die bisher bekannten Jugendstadien der zweiflugeligen Insekten Pt. 1. Berlin. 184 p.
___. **1950.** Die larvenformen der Dipteren. Eine Übersicht über die bisher bekannten Jugendstadien der zweiflugeligen Insekten Pt. 2. Berlin. 458 p.

Hindes, W.D. 1929. Notes on the genus *Polypedilum* Kieffer (Diptera, Chironomidae). The Entomologist's Monthly Magazine 65: 178-179.
Hinton, H.E. 1951. A new chironomid from Africa, the larva of which can be dehydrated without injury. Procedings of the Zoological Society of London 121: 371-380.
___. **1960.** Cryptobiosis in the larva of *Polypedilum vanderplanki* Hint. (Chironomidae). Journal of Insect Physiology 5: 286-300.
Hirvenoja, Mauri. 1962. Zur Kenntnis der Gattung *Polypedilum* Kieff. (Dipt., Chironomidae). Annales Entomologici Fennici 28: 127-136.
___. **1973.** Revision der Gattung *Cricotopus* van der Wulp und ihrer Verwandten (Diptera, Chironomidae). Annales Zoologici Fennici 10: 1-363.
Hudson, Patrick L., David R. Lenat, Broughton A. Caldwell, and David R. Smith. 1990. Chironomidae of the southeastern United States: A checklist of species and notes on biology, distribution and habitat. Fish and Wildlife Research 7, United States Department of the Interior. Washington, D.C. 46 p.
Jankovic, Mirjana. 1973. Die Chironomidenfauna aus dem Periphyton in der jugoslawischen Donaustrecke. Archive für Hydrobiologie, Supplementband 44: 249-257.
Johannsen, Oskar A. 1905. Aquatic nematocerous Diptera II. Pages 76-315 *in* Mayflies and Midges of New York by James G. Needham, Kenneth J. Morton, and Oskar A Johannsen. New York State Museum Bulletin 86: 7-352.
___. **1907.** Appendix. Some new species of Kansas Chironomidae. Pages 109-112 *in* Some results of desultory collecting of insects in Kansas and Colorado by E.S. Tucker. Kansas University Science Bulletin 4: 51-112.
___. **1908.** New North American Chironomidae. Pages 264-285 *in* 23rd report of the state entomologist on injurious and other insects of the state of New York 1907 by E.P. Felt. New York State Museum Bulletin 124: 5-541.
___. **1926.** Notes on the synonymy of some New York State Chironomidae. Journal of the New York Entomological Society 34: 273-277.
___. **1937.** Part IV. Chironomidae: Subfamily Chironominae. Pages 3-52 *in* Aquatic Diptera Parts IV and V by Oskar A. Johannsen and Lillian C. Thomsen. Cornell University Agricultural Experiment Station, Memoir 210: 80 p. + 12 plates. (1969 unabridged facsimile reprint from Entomological Reprint Specialists).
Jonasson, Petur M. 1948. Quantitative studies of the bottom fauna. Pages 204-284 *in* Biological studies on the River Susaa by Kaj Berg. Folia Limnologica scandinavica 4: 1-318.
Kertesz, Kalman. 1902. Catalogus dipterorum huicusque descriptorum. Vol. 1. Leipzig, Budapest. 339 p.
Kieffer, Jean-Jacques. 1906. Diptera. Fam. Chironomidae. Fasc. 42 of Genera insectorum. P. Wytsman, ed. Brussels. 78 p.
___. **1911.** Nouvelles descriptions de Chironomides obtenus d'eclosion. Bulletin de la Societe d'Histoire Naturelle de Metz 27: 1-60.
___. **1912.** Tendipedidae. H. Sauter's Formosa-Ausbeute. Tendipedidae (Chironomidae) [Dipt.]. Supplementa Entomologica 1: 27-43.
___. **1913.** Nouveaux Chironomides (Tendipedides) d'Allemagne. Bulletin de la Société d'Histoire Naturelle de Metz 28: 7-35.
___. **1915.** Neue Chironomiden aus Mitteleuropa. Brotéria, Série Zoológica 13: 65-87. [not seen]
___. **1920.** Tableau synoptique suivant des Chironomides paléarctiques appartenant aux genres *Polypedilum* et *Limnochironomus*. Annales de la Société Scientifique de Bruxelles 39: 159-167.
___. **1921a.** Chironomides de Courlande. Annales de Société Scientifique de Bruxelles 40: 275-298.
___. **1921b.** Nouveaux représentants du genre *Polypedilum* (Chironomides). Annales de Société Scientifique de Bruxelles 41: 97-101.
___. **1921c.** Chironomides nouveaux ou peu connus de la région paléarctique. Bulletin de la Société d'Histoire Naturelle de la Moselle, Metz 29: 51-109.
___. **1922a.** Nouveaux Chironomides a larves aquatiques. Annales de Société Scientifique de Bruxelles 41: 355-366.
___. **1922b.** Chironomides nouveaux ou peu connus de la région paléarctique. Annales de Société Scientifique de Bruxelles 42: 71-128; 138-180.
Knight, Kenneth L. and Jean L. Laffoon. 1971. A mosquito taxonomic glossary V. Abdomen (except female genitalia). Mosquito Systematics Newsletter 3: 8-24.
Kruseman, Gideon. 1933. Tendipedidae Neerlandicae I: Genus *Tendipes* cum generibus finitimis. Tijdschrift voor Entomologie 76: 119-216.

Kugler, Jehoshua and Friedrich Reiss. 1973. Die triangularis-Gruppe der Gattung *Tanytarsus* v.d.W. (Chironomidae, Diptera). Entomologisk tidskrift 94: 59-82.

Laffoon, Jean L. and Kenneth L. Knight. 1973. A mosquito taxonomy glossary IX. The larval cranium. Mosquito Systematics 5: 31-96.

Langton, Peter H. 1984. A key to pupal exuviae of British Chironomidae. Peter H. Langton, Huntingdon, Cambridgeshire. 324 p.

___. **1991.** A key to pupal exuviae of west Palaearctic Chironomidae. Peter H. Langton, Huntingdon, Cambridgeshire. 386 p. [not seen]

___. **1994.** If not "filaments," then what? Chironomus 6: 9. [not seen]

Leathers, Adelbert L. 1922a. *Chironomus braseniae*, new species (Dip., Chironomidae). Entomological News 33: 8.

___. **1922b.** Ecological study of aquatic midges and some related insects with special reference to feeding habits. United States Bulletin of the Bureau of Fisheries 38: 1-61.

Lehmann, Jens. 1971. Die Chironomiden der Fulda (Systematische, okologische und faunistische Untersuchungen). Archiv für Hydrobiologie, Supplementband 37: 466-555.

Lenz, Friedrich. 1921. Chironomidenpuppen und -larven. Bestimmungstabellen. Deutsche Entomologische Zeitschrift 1921: 148-162.

___. **1923.** Die Vertikalverteilung der Chironomiden im eutrophen See. Verhandlungen der Internationalen Vereinigung für Theoretische und Angewandte Limnologie 1: 144-167.

___. **1941.** Die Jugendstadien der Sectio Chironomarie (Tendipedini) connectentes (Subf. Chironominae = Tendipedinae). Zusammenfassung und Revision. Archiv für Hydrobiologie 38: 1-69.

___. **1962.** Tendipedidae (Chironomidae). b) Subfamilie Tendipedinae (Chironominae). Gattung *Polypedilum* Kieff. and *Pentapedilum* Kieff. Pages 237-245 *in* Die Fliegen de palaearktischen Region. Erwin B. Lindner, ed. 13c, Lieferung 222: 233-260.

Lindegaard-Peterson, Claus. 1972. An ecological investigation of the Chironomidae (Diptera) from a Danish lowland stream (Linding A). Archiv für Hydrobiologie 69: 465-507.

Malloch, John R. 1915. The Chironomidae, or midges, of Illinois, with particular reference to the species occurring in the Illinois River. Bulletin of the Illinois State Laboratory of Natural History 10: 275-543.

Matsuda, Ryuichi. 1965. Morphology and evolution of the insect head. Memoirs of the American Entomological Institute, no. 4. 334 p.

McGaha, Y.J. 1952. The limnological relations of insects to certain aquatic flowering plants. Transactions of the American Microscopical Society 71: 355-381.

Meigen, Johann W. 1804. Klassifikazion und Beschreibung der europäischen zweiflügeligen Insekten. (Diptera Linn.). Braunschweig, 1, Abt. 1, I-xxviii+152 p.; Abt. 2 vi + 161 p. (Numbered 153-314).

___. **1818.** Systematische Beschreibung der bekannten europäischen, zweiflügeligen Insekten. Aachen, 1, xxxvi + 332 p.

___. **1830.** Systematische Beschreibung der bekannten europäischen, zweiflügeligen Insekten. Hamm., 6, xi + 401 p.

___. **1838.** Systematische Beschreibung der bekannten europäischen zweiflügeligen Insekten (Supplementary volume). Hamm., 7. xii + 434 p.

Meuche, Alfred. 1939. Die Fauna im Algenbewuchs. Nach Untersuchungen im Litoral ostholsteinischer Seen. Archiv für Hydrobiologie 34: 349-520.

Mik, Josef. 1884. Nachträge zu Schiners "Fauna Austriaca" (Diptera). Wiener Entomologische Zeitung 3: 201-206.

Miller, P.L. 1969. On the occurrence and some characteristics of *Cyrtopus fastuosus* Bigot (Dipt., Stratiomyidae) and *Polypedilum* sp. (Dipt., Chironomidae) from temporary habitats in western Nigeria. Entomologist's Monthly Magazine 105: 233-238.

Miller, Richard B. 1941. A contribution to the ecology of the Chironomidae of Costello Lake, Algonquin Park, Ontario. University of Toronto Studies, Biological Series 49: 1-63.

Mitchell, J.D. 1908. Descriptions of nine new species of gnats. Journal of the New York Entomological Society 16: 7-14.

Moller Pillot, Henk K.M. 1978-1979. De larven der Nederlandse Chironomidae (Diptera). Nederlandse Faunistische Mededelingen, 1. 277 p.

Morgan, N.C. and Alexandra B. Waddell. 1961. Diurnal variation in the emergence of some aquatic insects. Transactions of the Royal Entomological Society of London 113: 123-137.

Mozley, Samuel C. 1970. Morphology and ecology of the larva of *Trissocladius grandis* (Kieffer) (Diptera, Chironomidae), a common species in the lakes and rivers of northern Europe. Archiv für Hydrobiologie 67: 433-451.

___**. 1971.** Maxillary and premental patterns in Chironominae and Orthocladiinae (Diptera: Chironomidae). The Canadian Entomologist 103: 298-305.

Muttkowski, R.A. 1918. The fauna of Lake Mendota, a qualitative and quantitative survey with special reference to the insects. Transaction of the Wisconsin Academy of Sciences, Arts and Letters 19: 374-482.

Mundie, J.H. 1957. The ecology of Chironomidae in storage reservoirs. Transactions of the Royal Entomological Society of London 109: 149-232.

Niitsuma, Hiromi. 1992. The *Polypedilum convictum* species group (Diptera, Chironomidae) from Japan, with descriptions of two new species. Japanese Journal of Entomology 60: 693-706.

Oliver, Donald R., and Mary E. Dillon. 1989. 2. The adult males of Chironomidae (Diptera) of the Holarctic region—key to subfamilies. Pages 11-15 *in* Chironomidae of the Holarctic region keys and diagnoses. Part 3. Adult males. Torgny Wiederholm, ed. Entomological scandinavica Supplement 34. 532 p.

___**. 1994.** Corrections and additions to "A catalog of Nearctic Chironomidae." Proceedings of the Entomological Society of Washington 96: 8-10.

Oliver, Donald R., Mary E. Dillon, and Peter S. Cranston. 1990. A catalog of Nearctic Chironomidae. Research Branch, Agriculture Canada. Publication 1857/B. 89 p.

Oyewo, Emmanuel Adeoye and Ole A. Sæther. 1998. Revision of Afrotropical *Polypedilum* Kieffer subgenus *Uresipedilum* Sasa et Kikuchi, 1995 (Diptera: Chironomidae), with a review of the subgenus. Annals of Limnology 34: 315-362.

Pagast, Felix. 1933. Chironomidenstudien. Stettiner Entomologische Zeitung 94: 286-300.

___**. 1947.** Systematik und Verbreitung der um die Gattung *Diamesa* gruppierten Chironomiden. Archiv für Hydrobiologie 41: 435-596.

Paine, George H. and Arden R. Gaufin. 1956. Aquatic Diptera as indicators of pollution in a midwestern stream. The Ohio Journal of Science 56: 291-304.

Palmén, Ernst. 1955. Diel periodicity of pupal emergence in natural populations of some chironomids (Diptera). Annales Zoologici Societatis Zoologicæ Botanicæ Fennicæ 'Vanamo' 17: 1-30.

___**. 1958a.** Periodic emergence in some North European chironomids. Internationale Vereinigung für Theoretisch und Angewandte Limnologie, Verhandlungen 13: 817-822.

___**. 1958b.** Diel periodicity of pupal emergence of some north European chironomids. Proceedings of the 10th International Congress of Entomology, Montreal, 1956. 2: 219-224.

Pettit, R.H. 1900. A leaf-miner, *Cheironomus* (*sic*) sp., in water lilies. Michigan Academy of Science 1: 110-111.

Pinder, L. Clive V. 1974. The Chironomidae of a small chalk-stream in southern England. Entomologisk tidskrift, Supplementum 95: 195-202.

___**. 1983.** 1. The larvae of Chironomidae (Diptera) of the Holarctic region—Introduction. Pages 7-10 *in* Chironomidae of the Holarctic region—keys and diagnoses. Part 1. Larvae. Torgny Wiederholm, ed. Entomologica scandinavica Supplement 19. 457 p.

___**. 1989.** 1. The adult males of Chironomidae (Diptera) of the Holarctic region—Introduction. Pages 5-9 *in* Chironomidae of the Holarctic region—keys and diagnoses. Part 3. Adult males. Torgny Wiederholm, ed. Entomologica scandinavica Supplement 34. 532 p.

Pinder, L. Clive V. and Friedrich Reiss. 1983. 10. The larvae of Chironomidae (Diptera) of the Holarctic region—Keys and diagnosis. Pages 293-435 *in* Chironomidae of the Holarctic region—keys and diagnoses. Part 1. Larvae. Torgny Wiederholm, ed. Entomologica scandinavica Supplement 19. 457 p.

___**. 1986.** 10. The pupae of Chironomidae (Diptera) of the Holarctic region—Keys and diagnosis. Pages 299-456 *in* Chironomidae of the Holarctic region—keys and diagnoses. Part 2. Pupae. Torgny Wiederholm, ed. Entomologica scandinavica Supplement 28. 482 p.

Prashed, Baini. 1918. The anatomy of a chironomid larva of the genus *Polypedilum*. Record of the Indian Museum 14: 71-74.

Reiss, Friedrich. 1968. Ökologische und systematische Untersuchungen an Chironomiden (Diptera) des Bodensees. Ein Beitrag zur lakustrischen Chironomidenfauna des nördlichen Alpenvorlandes. Archiv für Hydrobiologie 64: 176-323.

___**. 1969.** *Krenopsectra fallax* gen. n. sp. n. (Diptera, Chironomidae) aus den Alpen und den Pyrenaen. Annales Zoologici Fennici 6: 435-442.

Roback, Selwyn S. 1953. Savannah river tendipedid larvae [Diptera: Tendipedidae (= Chironomidae)]. Proceedings of the Academy of Natural Sciences of Philadelphia 105: 91-132.

___. **1957.** The immature tendipedids of the Philadelphia area (Diptera: Tendipedidae). Monographs of the Academy of Natural Sciences of Philadelphia 9: 1-152.

Rossaro, Bruno. 1985. Revision of the genus *Polypedilum* Kieffer, 1912. I. Key to adults, pupae and larvae of the species known to occur in Italy. Memorie della Società Entomologica Italiana, Genova 62/63 (1983-1984): 3-23.

Sæther, Ole A. 1969. Some Nearctic Podonominae, Diamesinae, and Orthocladiinae. Fisheries Research Board of Canada, Bulletin 170: 1-154.

___. **1971.** Notes on general morphology and terminology of the Chironomidae (Diptera). The Canadian Entomologist 103: 1237-1260.

___. **1975.** Nearctic chironomids as indicators of lake topology. Internationale Vereinigung für Theoretische und Angewandte Limnologie, Verhandlungen 19: 3127-3133.

___. **1980.** Glossary of chironomid morphology terminology (Diptera: Chironomidae). Entomologica scandinavica Supplement 14. 51 p.

___. **1990.** A review of the genus *Limnophyes* Eaton from the Holarctic and Afrotropical regions (Diptera: Chironomidae, Orthocladiinae). Entomologica scandinavica Supplement 35. 139 p.

Sæther, Ole A. and Asbjørn Sundal. 1999. *Cerobregma*, a new subgenus of *Polypedilum* Kieffer, with a tentative phylogeny of subgenera and species groups within *Polypedilum* (Diptera: Chironomidae). Journal of the Kansas Entomological Society 71(3) 1998: 315-382.

Sandberg, Gosta. 1969. A quantitative study of chironomid distribution and emergence in Lake Erken. Archiv für Hydrobiologie, Supplementband 35: 119-201.

Sasa, Manabu and Mihoko Kikuchi. 1995. Chironomidae [Diptera] of Japan. University of Tokyo Press. 333 p.

Schiner, J. Rudolph. 1864. Fauna Austriaca. Die Fliegen (Diptera). Wien. 2: 574-622.

Schlee, Dieter. 1966. Präparation und Ermittlung von Messwerten an Chironomidae (Diptera). Gewässer und Abwässer 41/42: 169-193.

Snodgrass, Robert E. 1935. Principles of insect morphology. McGraw-Hill Book Co., Inc. New York and London. 667 p.

___. **1947.** The insect cranium and "epicranial suture." Smithsonian Miscellaneous Collections 107: 1-52.

___. **1959.** The anatomical life of the mosquito. Smithsonian Miscellaneous Collections 139: 1-87.

___. **1960.** Facts and theories concerning the insect head. Smithsonian Miscellaneous Collections 142: 1-61.

Soponis, Annelle R. 1983. Emergence of *Polypedilum* (Chironomidae) in a sand-bottomed stream of northern Florida. Memoirs of the American Entomological Society 34: 309-313.

Soponis, Annelle R. and Constance L. Russell. 1982. Identification of instars and species in some larval *Polypedilum* (*Polypedilum*) (Diptera: Chironomidae). Hydrobiologia 94: 25-32.

Soponis, Annelle R. and Karl W. Simpson. 1992. *Polypedilum digitifer* Townes and *Polypedilum griseopunctatum* (Malloch) (Diptera, Chironomidae): redescription of adult males with description and separation of the immature stages. Netherlands Journal of Aquatic Ecology 26: 203-213.

Strenzke, Karl. 1959. Revision der Gattung *Chironomus* Meig. I. Die Imagines von 15 norddeutschen Arten und Unterarten. Archiv für Hydrobiologie 56: 1-42.

___. **1960.** Metamorphose und Verwandtschaftsbeziehungen der Gattung *Clunio* Hal. (Dipt.). Annales Zoologici Societatis Zoologicæ Botanicæ Fennicæ 'Vanamo' 22: 1-30.

Stuart, T.A. 1941. Chironomid larvae of the Millport shore pools. Transactions of the Royal Society of Edinburgh 60: 475-502.

Sublette, James E. 1957a. The immature stages and females of *Polypedilum* (*Polypedilum*) *obtusum* Townes (Tendipedidae: Diptera). Entomological News 68: 37-40.

___. **1957b.** The ecology of the macroscopic bottom fauna in Lake Texoma (Denison Reservoir), Oklahoma and Texas. The American Midland Naturalist 57: 371-402.

___. **1960.** Chironomid midges of California. I, Chironominae, exclusive of Tanytarsini (=Calopsectrini). Proceedings of the United States National Museum 112: 197-226.

___. **1964.** Chironomidae (Diptera) of Louisiana I. Systematics and immature stages of some lentic chironomids of west-central Louisiana. Tulane Studies in Zoology 11: 109-150.

___. **1966.** Type specimens of Chironomidae (Diptera) in the American Museum of Natural History. Journal of the Kansas Entomological Society 39: 1-32.

Sublette, James E. and Manabu Sasa. 1994. Chironomidae collected in Onchocerciasis endemic areas of Guatemala (Insecta, Diptera). Spixiana Supplement 20: 1-60.

Sublette, James E. and Mary S. Sublette. 1965. Family Chironomidae (Tendipedidae). Pages 142-181 *in* A catalog of the Diptera of America North of Mexico by Alan Stone, *et al.* Agricultural Research Service, Agricultural Handbook No. 276. 1696 p.

___. **1973.** The morphology of *Glyptotendipes barbipes* (Staeger) (Diptera, Chironomidae). Studies in Natural Sciences, Portales, New Mexico volume 1, number 6. 81 p.

Tebo, Leonidas B., Jr. 1955. Bottom fauna of a shallow eutrophic lake, Lizard Lake, Pocahontas County, Iowa. The American Midland Naturalist 54: 89-103.

Thienemann, August. 1954. *Chironomus.* Leben, Verbreitung und wirtschaftliche Bedeutung der Chironomiden. Binnengewässer 20: 1-834.

Tokunaga, Masaaki. 1932. Morphological and biological studies on a new marine chironomid fly, *Pontomyia pacifica*, from Japan. Part 1. Morphology and taxonomy. Memoirs of the College of Agriculture, Kyoto University (Entomology Series no. 3) 19: 1-56.

___. **1964.** Diptera: Chironomidae. Pages 485-628 *in* Insects of Micronesia, Diptera. Bernice P. Bishop Museum 12 (5). 635 p.

Townes, Henry K., Jr. 1945. The nearctic species of Tendipedini (Diptera, Tendipedidae) (=Chironomidae). The American Midland Naturalist 34: 1-206.

___. **1952.** Tribe Tendipedini (=Chironomini). Pages 27-103 *in* Guide to the insects of Connecticut. Part VI. The Diptera or true flies of Connecticut. Fifth fascicle: Midges and gnats Tendipedini (Chironomidae) by Oskar A. Johannsen and Henry K.Townes, Jr. Connecticut State Geology and Natural History Survey, Bulletin 80. 255 p.

Walker, Francis. 1856. Diptera. Vol. 3 of Insecta Britannica by Francis Walker, H.T. Stainton, and S.J. Wilkinson. London, England. xxiv + 352 p.

Walley, G. Stuart. 1926. Four new Canadian Chironomidae. The Canadian Entomologist 58: 205-207.

Weber, Cornelius I., ed. 1973. Biological field and laboratory methods for measuring the quality of surface waters and effluents. National Environmental Research Center, Office of Research and Development, U.S. Environmental Protection Agency. EPA-670/4-73-001. Cincinnati, Ohio. [sections numbered separately].

Wurtz, Charles B. and Selwyn S. Roback. 1955. The invertebrate fauna of some Gulf Coast Rivers. Proceedings of the Academy of Natural Sciences of Philadelphia 107: 167-206.

___. **1877.** Diptera Neerlandica. De tweevleugelige insecten van Nederland. I. 497 p.

Zavřel, Jan. 1941. Vergleichend—morphologische Untersuchungen an den Podomominenlarven (Diptera, Chironomidae). I. Labrum und Präemandibeln. Zoologischer Anzeiger 134: 105-115.

___. **1942.** Polypodie pakomá ich kukel. Polypodie der Chironomiden-Puppen. Acta Societatis Scientarum Naturalium Moravicae 14: 1-40.

Zetterstedt, Johan W. 1838. Dipterologis scandinaviae. Sect. 3: Diptera, 477-868.

___. **1850.** Diptera Scandinaviae disposita et discripta. IX: 3367-3710. Lundae.

Index to Taxa

(Subgenera are within parenthesis; names in **bold** are new species; names in *italics* are synonyms; page numbers in **bold** indicate descriptions included in this book; page numbers in *italics* indicate figures)